IN SEARCH OF
THE
BEGINNING

A Seeker's
Journey to
the Origin of
the Universe,
Life, and Man

ATHENIANS QUESTIONS (P12)

FOUR TRUTH MONITORS (P42)

Def. "WORKING" P43

IN SEARCH OF
THE
BEGINNING

A Seeker's
Journey to
the Origin of
the Universe,
Life, and Man

DEAN DAVIS

Pleasant Word
A Division of WINEPRESS PUBLISHING

Pleasant Word (a division of WinePress Publishing, PO Box 428, Enumclaw, WA 98022) functions only as book publisher. As such, the ultimate design, content, editorial accuracy, and views expressed or implied in this work are those of the author.

Unless otherwise noted, all Scriptures are taken from the New King James Version of the Bible.

Scripture references marked KJV are taken from the King James Version of the Bible.

Scripture references marked NASB are taken from the New American Standard Bible, © 1960, 1963, 1968, 1971, 1972, 1973, 1975, 1977 by The Lockman Foundation. Used by permission.

ISBN 1-4141-0371-9
Library of Congress Catalog Card Number: 2004195526

Dedication

This book is gratefully dedicated to the unknown god—and to seekers everywhere who hope to meet him at the beginning.

I want to know how God created the world. I am not much interested in this or that phenomenon in the spectrum of this or that element. I want to know His thoughts. The rest are details.

— Albert Einstein

Seek and ye shall find.

— Jesus of Nazareth

Table of Contents

Acknowledgments

There is much doubt attached to modern cosmology, but of one thing we may be certain: scientific laymen trained in the humanities rarely write books about the origin of the universe, life and man. That I have dared to do so is testimony to the invaluable help supplied by many dear friends, colleagues and loved ones.

Many thanks to Tammy Hopf and Tim Noreen, my liaisons, and to the entire staff at Pleasant Word. This was a challenging project, one which their professionalism, competence, and good humor transformed into an adventure and a delight. Lord willing, I'll be back!

My indebtedness to professional scientists who have written for laymen like me will become crystal clear in the pages ahead. Here, however, I would like especially to thank my new friend Dr. John Byl, who patiently answered my frequent questions, reviewed the manuscript, and offered many valuable suggestions. If my own book somehow succeeded in alerting curious seekers to the excellencies of his, I would be pleased indeed.

Thanks also to my old and faithful friend Steve Carver, who worked his usual magic on the computer to create the diagrams and tables you will find in the text.

I owe special thanks to my father, Don Davis, who has faithfully supported me in this and other literary ventures. Dad, you have shown yourself a true patron—in every sense of that rich and venerable word.

Finally, heartfelt thanks to my dear wife, Linda. Honey, I will never forget how you pulled me through those two terrible midnight hours when I thought I had erased four-fifths of my book from the new computer. It was only one of your many expressions of love and support. No, I will never forget.

The Test, the Teacher, and the Beginning

This is a book about the great themes of cosmology—the origin, structure, purpose, and destiny of the universe. But it is especially about the one theme that serves as seed for all the rest: the beginning—the origin of the universe, life, and man.

It is also a book written specially for seekers—people who believe, or at least suspect, that there is more to the universe than the universe; people who sense that there is an ultimate spiritual reality back of all things, a reality about which they are curious to know more. If you fall into this category, you will find in the pages ahead something you may have been looking for: a meditation on the cosmos that is attentive not only to natural science, but to philosophy and theology as well.

Finally, this is, as it were, a book within a book. It was first conceived as a humble chapter, part of a much larger work in progress called *The Test: A Seeker's Journey Towards the Meaning of Life*. In time, the one chapter became two, two became four, and four became the volume you now hold in your hands. What produced this literary cell division, I do not quite know. Perhaps it was my keen interest in the subject, or its boundless philosophical importance, or its rich (and sometimes maddening) complexity, or the controversy that has often surrounded it, never more so than today.

In any case, my "chapter" has at last reached full stature and I am eager to introduce it to you. But for the meeting to go well, it is important that you know a little more about its mother. Permit me, then, to break with literary protocol by devoting the lion's share of a lengthy introduction, not to the present book, but to a summary of the book from which it sprang —*The Test*. If these preliminaries seem to meander too far from your interest, you may, of course, skip over to Chapter 1 and plunge in. But I counsel against it. Better, I think, to hike awhile here in the lowlands. Having done so, you will find yourself in far better shape for a challenging climb that will follow soon enough.

Life: A Mess or a Test?

The Test is structured as a spiritual journey in which the first step is to experience a fundamental change in perspective. To this end, I begin by reminding my readers of a fact most of us know very well but sometimes try to avoid: that each of us has *a heart full of questions*. Here I especially have in mind what are sometimes called life's "ultimate questions"—the distinctly religious and philosophical problems that have ever occupied the deepest thoughts and concerns of the human race.

In *The Test*, I identify an irreducible core of nine such questions:

1. What is the ultimate reality?
2. What is the origin of the universe, life, and man?
3. What, if anything, went wrong? Why is there evil, suffering, and death in the world?
4. What, if anything, can be done about them?
5. What is the meaning of life?
6. How should we live?
7. What happens when we die?

8. Where is (cosmic) history heading?

9. How can we find trustworthy answers to the questions of life?

Pondering these questions, we soon realize that they display several interesting and significant characteristics.

First, they are *universal*. As a study of world religion and philosophy will show, people of all times and all places have wrestled with them. This may not seem surprising, but in a way it is. Why should all people, and not just some, think about these questions? Could it be that the questions somehow belong to human nature itself? And if so, how did they get there?

Second, they are *existentially urgent*. By this I mean that we care, and care deeply, about finding the answers. Indeed, I think most of us would admit that our own sense of personal well being depends heavily upon discovering the truth about one or more of the questions of life.

Finally, these questions are closely related to what philosophers call our *worldview*. In fact, a worldview may be precisely defined as *a way of looking at reality as a whole, based upon a particular set of answers to the questions of life*. The prominent place of religion and philosophy in human experience testifies to the fact that most of us actively seek a worldview. Furthermore, any old worldview will not do: what we really want is *the one true worldview*. Question by question, answer by answer, step by step, we would ascend to that intellectual vantage point from which alone we can at last see all of reality as it really is. And we do this not only because we desire to see reality, but also because we desire to *relate* to it as it really is. Deep in our hearts we sense that finding the true worldview is a very special kind—indeed, the ultimate kind—of coming home.

But there is a problem. For though we all have a heart full of questions, we do not all have a heart full of answers. The answers we want and need are not self-evident, else we would not be seeking them. To use a humble illustration, we find that our hearts are rather like an incomplete wooden jigsaw puzzle: the questions of life are the spaces, but the answers that fit into the spaces are nowhere in sight.

And so, in an effort to fill the voids, some among us have stepped forward with answers, often quite dogmatically.

Consider, for example, the question of the ultimate reality. Some have claimed that the ultimate reality is energy/matter, so that all the things we call things are simply an embodiment of this one primordial substance. This is the view of the philosophical *naturalist*.

Others, however, argue that the ultimate reality is an impersonal divine mind that, in the ongoing experience of billions of sentient beings, has somehow slipped into a cosmic dream. We humans think that we are souls, inhabiting bodies, living in a real material world. But that is an illusion. The truth is that we are simply tiny bubbles of consciousness, arising in the infinite ocean of Big Mind. Here is the view of the *pantheist*.

Meanwhile, still others assert that the ultimate reality is an infinite personal god—a god who created and sustains all things but remains metaphysically separate from them. This is the view of the *theist*.[1]

We see, then, that people hold different views of the ultimate reality. And what is true for the question about ultimate reality is true for the other questions of life as well: there are always several possible answers, each answer differing significantly from the others.

Thus, poor seekers after truth are in a quandary. They have a heart full of existentially urgent questions but, when they look around for truth, they find *a world full of contradictory answers*! This situation is scandalous. Deep down we feel it ought not to be. But it is—and never more so than today, when sophisticated media have brought every conceivable answer and worldview directly to our doorsteps. How, then, are we to understand and respond to a philosophical situation such as this?

I see two possibilities, two perspectives.

First, there is what I will call *the mess perspective*. Sometimes referred to as *postmodernism* (and previously called skepticism), this view holds that life—religiously and philosophically speaking—is a mess. In other words, there is no such thing as objective truth or absolute values. The fact that different worldviews *do* exist proves that no true worldview *can* exist. We are all stuck in our heads. The way we see things is relative to our language, history, imagination, culture, and perhaps even to our biology, all of which differ from place to place and time to time. Accordingly, religions and philosophies must be seen as stories or "meta-narratives"—culturally determined word pictures designed to help people get a handle on the world around them. In the past such handles may have had some survival value. But in today's shrinking world we dare not take them too seriously. Even if we cannot learn truth, at least we can learn to get along. Let us, therefore, come of age. Let us abandon our quest for true answers to the questions of life. Let us surrender our hopes of ever finding the true worldview. Let us simply live and let live, tolerating and respecting each other's stories. In sum, however disappointing or frustrating it may seem, let all the family of man accept and get used to the fact that life is mess.

Does this take on the human condition depress you? If so, good! That means you are an excellent candidate for a second and far more hopeful point of view. I call it *the test perspective*.

According to this view, life is a test placed before us by an unknown god. He himself has put into our hearts the questions of life, as well as an abiding hope of finding the answers. But for wise reasons he has not made the answers self-evident. Moreover, he has also allowed a certain amount of religious and philosophical error to creep into his world. Thus, he has set the stage. What will his human creatures do now? Will they listen to their hearts and begin sorting through the various philosophical options till they find the truth? Or will they use the existence of options as an excuse not to seek truth but to do what they want? As each of us decides, this god is watching. If we seek, he will help us. If we find, he will reward us. The test is on. Our part is as simple as it is important: we must love the truth enough to seek it, and we must seek it until we find it.

Now most folks would agree that this is indeed a more hopeful way of looking at life. They would like to know, however, if there are any good reasons to believe it is true? Because I believe there are, I will touch on a few of the most important here.

First, there is the lesson of natural hunger and thirst. In the natural world there is an objective reality that corresponds to our hunger: food. There is also an objective reality that corresponds to our thirst: drink. Interestingly, we often have to seek out food and drink, and can usually find them if we want them badly enough. Do these simple facts of daily life have a message for us? Is the natural world teaching us something important about the spiritual? Does our hunger for truth also correspond to an objective reality? Does it imply that truth exists? And does it imply that truth will supply both nourishment and pleasure if and when we find it?[2]

Second, there is the amazing makeup of the human mind. How is it that we are all endowed with intellect, intuition, conscience, language, and curiosity? How is it that we often focus these faculties intently on the questions of life? And how is it that we are surrounded by other minds with whom we may readily think about those questions? Viewed from one angle, it certainly looks as if we humans have been *equipped* for a search for truth. The tools are in us and around us. Our part, it would appear, is simply to use them.

Third, there is what I like to call "the manageable messiness" of the religious/philosophical world. The idea here is that our spiritual condition is not nearly so messy as our postmodern friends would have us believe.

For example, we have already seen that the questions of life are relatively few—about nine in number. What's more, they are easy enough to understand. Children and youth ask them all the time—even if we adults cannot answer them all the time.

Also, the possible answers to the questions are few and easy enough to understand. For example, to the question, "What happens when we die?" religion and philosophy repeatedly return to three basic options: the lights go out (the view of the naturalist), the soul reincarnates (the view of the pantheist), or the soul goes immediately to Heaven, purgatory or hell (the view of the theist). We may not like some of these answers or find them equally plausible, but no one can say they are too numerous or too difficult to comprehend.

Of special interest is the fact that the possible worldviews are *very* few, and also relatively easy to understand. In *The Test* I argue that there are really only three basic worldviews (naturalism, pantheism and theism) and that the shape of each one flows logically from their respective understandings of the ultimate reality. Yes, some confusion arises because there are quite a number of spokesmen for each worldview, each with his or her unique twist. But in the end, nearly every religion or philosophy is easily identifiable as a species of naturalism, pantheism, or theism. Please reflect on this carefully; the more you do, the more you will find the paucity of worldviews to be a richly significant and deeply heartening fact of philosophical life.

Now in light of all this evidence I, for one, must conclude that mankind does not live in the midst of a philosophical chaos after all. To the contrary, it appears that human existence—though philosophically burdensome—is nevertheless *ordered*. Could it be, then, that just as we live in a natural order and a moral order, so too we live in a *probationary order*? Could it be that a rational supreme being—an unknown god—really is putting us to the test?

THE PROBATIONARY ORDER

I. A SPIRITUALLY EQUIPPED HUMAN BEING...
- A. Intuition
- B. Reason
- C. Language
- D. Community and communication
- E. Conscience
- F. Hope

II. CHALLENGED BY THE QUESTIONS OF LIFE...
- A. Innate questions
- B. Curiosity about the answers
- C. Existential urgency

III. IGNORANT OF THE ANSWERS...
- A. The answers are not within
- B. To find them we must look without

IV. SITUATED IN A MANAGABLY MESSY RELIGIOUS AND PHILOSOPHICAL WORLD...
- A. The questions are few and easy to understand
- B. The possible answers are few and easy to understand
- C. The possible worldviews are very few and easy to understand

V. AND FREE TO SEEK THE TRUTH OR NOT!

Let the Test begin!

The thought that we are on divine probation can be intimidating, for it is only natural to wonder what will happen if we fail the test. On the other hand, the same thought can be profoundly encouraging, for it means that while our life may indeed be difficult, it is definitely not absurd. As a matter of fact, from within the test perspective a previously messy life is suddenly revolutionized—charged with new meaning, adventure, and hope. The meaning of life—or at least its highest meaning—would be to seek, find, and respond to the truth. The adventure would be to face and overcome every obstacle standing in the way. And the hope would be not only to find the truth, but—just perhaps—the divine Tester as well!

Hints of a Heavenly Hope

Spiritually hungry souls—and especially souls entangled in post-modern despair—would doubtless rejoice to know that life is a test. But are there any further reasons to believe that an unknown god exists and that he has indeed situated us in a probationary order? Again, I believe there are. And interestingly enough, the evidence largely consists of two more orders. Let me touch briefly on both.

First, there is the natural order: the universe, life, and man. Under close examination we find that many characteristics of this order not only point to an unknown god but also reveal to us something of his nature.

For example, the universe and all things in it are marked by *dependency.* Their existence, cohesiveness, and motions—as well as the mystery of "life" within its many life forms—all seem to depend upon a power that is beyond themselves. But what, or who, is that power?

Again, the physical universe is marked by *order*. From the tiniest atoms to the largest clusters of galaxies, nearly all things display structure, complexity, and beauty. And this is true not only of the forms of things, but of their motions, behaviors, and relationships. It is counterintuitive in the extreme to hold that "nature" produced an asparagus fern, or a hummingbird's feathers, or a human brain, or an ecosystem, or a solar system, or the cosmos itself—all by accident.

Inescapably, order in the universe evokes within us an awareness of a divine orderer—an intelligent, powerful, and profoundly artistic unknown god.

Then, too, there is the _goodness_ of the natural order—the tendency of all things not only to sustain life but also to contribute to its pleasure. Think, for example, of the sun, the air, the soil, the rains, the abundance of delicious foods in the world, and of all the materials suitable for building shelters or clothing bodies. Think of the hidden powers and principles of nature—mechanical, chemical, electromagnetic or nuclear—and how, by technological advance, they have all enriched our lives.

Yes, there is natural evil in the world: sickness, injury, famine, pestilence, earthquake, hurricane, and more. Such brute facts are problematic for any worldview, causing us to ask the third question of life, "What (if anything) went wrong?" It must be observed, however, that as a general rule goodness predominates in human experience. Moreover, goodness, rather than its opposite, is our instinctive expectation. Rarely does anyone ask, "What went right? Why is there so much goodness, pleasure, and life in the world?" So here—in nature's goodness and in our expectation of it—we again catch a glimpse of the unknown god, a good god who delights in giving good things to all living beings, and especially to the sons and daughters of men.

Summing up, then, we find that the dependency, order, and goodness of the natural world all work together to unveil to human hearts an unknown god—a god who is personal, powerful, rational, and good.

Second, there is the moral order. Unlike the natural order, this order is spiritual rather than physical, invisible rather than visible. Nevertheless, we are well aware of its existence and of its mighty power to influence our lives for good or ill.

The several elements of the moral order press themselves upon our consciousness daily. We all know, for example, that there are certain universal moral laws: we ought not to commit murder, steal, lie, etc. Rather, we ought to love, serve, be faithful, courageous, industrious, etc. By and large, all peoples agree about the content of the _moral law_, as any survey of world religion and philosophy will show. And even if they do not, this need not mean that the moral law does not exist, only that its hold upon some of us has been weakened—perhaps even dangerously so.

Next, there is *moral obligation*, an objective spiritual reality which somehow makes itself known to the subjective faculty we call conscience. Together with conscience, it speaks to us inwardly, urging us to live up to the moral law or to reconcile ourselves to it when we fall short. Moral obligation may be invisible, but millions will testify that it is as real as any mountain they have ever climbed.

Finally, there is *the law of moral cause and effect*. Our innate awareness of this law assures us that good will ultimately triumph over evil; that we will always reap what we sow; and that righteousness will bring reward and evil will bring retribution, if not in this life, then surely in the next.

Again, the moral order is spiritual rather than physical, but no less real or objective than the natural world itself. Like the wind, we cannot see it, but we can see its effects. Every day we observe people relating to it—striving to honor it, warring against it, stumbling over it, longing to be reconciled to it. It is just as pervasive, complex, powerful, and beautiful as the natural order. And like the natural order, it too manifests design and points to a person with a purpose. It too reveals a personal god who created it and sustains it. Here, however, we learn something different about this god: that he is a holy, sovereign, and righteous judge—and that he would have us live before him accordingly.

The natural and moral orders are, then, two powerful "hints of a heavenly hope," solid evidence for the existence of an unknown god. I say "unknown," yet because of these orders it appears we can actually know quite a bit about him: that he is personal, powerful, intelligent, wise, artistic, good, holy, sovereign, a respecter of our freedom, and a rewarder of those who use it well.

And there is one thing more: he certainly seems to enjoy creating orders!

Could it be, then, that there really is a probationary order, and that the god of the natural and moral orders is its author? With so much evidence for an orderly god before our eyes, this would certainly seem to be the case.

In Search of the Teacher

Now suppose that in contemplating these three orders someone awakens to the existence of an unknown god who is holding them together. Suppose he concludes that life is—or is very likely—a test. Suppose he decides to search out the answers to all the questions of life, and to learn all he can about the divine Tester. What then? How, practically speaking, is he to proceed?

To begin with, he should proceed by rejoicing, for a seeker has been born and a great journey—with great promise of great reward—is about to begin. If the test perspective is true, such things are a joy not only to man, but also to the unknown god himself.

But after the rejoicing, then what? What first baby steps is the new-born seeker to take?

In a sense, the answer to that question is already within him. For is it not the case that all of us, even from our childhood, seek truth at the feet of a teacher? Perhaps we turn to a parent, or to a pastor, or to a trusted professor. Whatever the case, it seems that we humans are "wired" to look for the truth outside of ourselves; to look for an authoritative "someone" from whom we can hear the special words that we know will bring us life.

Interestingly, this inclination makes excellent sense from within the test perspective. If we really believe that life is a test, then we know—or at least strongly suspect—that the divine Tester is on our side. But if he is on our side, then surely he *must* have made a provision for us to find the answers we need in order to pass his test. In other words, he must have sent us some kind of teacher, or at least be planning to in days ahead. The seeker's next step, then, is to begin looking for what I will call *god's appointed Teacher*—the person or group of persons authorized by the unknown god to bring us his true answers to the questions of life.

Now let us assume for the moment that such a teacher has already come into the world. How shall we find him? And how shall we be certain that we have found him when we do? In asking these questions, the seeker's search begins.

As a rule, it also begins with some dead ends.

Soon enough, for example, our seeker friend will realize that *nature* is not god's appointed Teacher. Nature, as we just saw, does indeed tell us a few things about the unknown god, but not nearly enough. Nature does not tell us god's name (if he has a name). It does not tell us all we want to know about his character, or his plans for the cosmos. It does not tell us how evil entered the universe or if and how it will be removed, etc. And what is true of the natural order is true of the moral order as well. Neither is god's appointed Teacher, for neither fully discloses to us the answers to the questions of life.

Similarly, it will not be long before the seeker realizes that *natural science* is not the Teacher. This only makes sense, since natural science is limited to the study of nature—the physical world—whereas the questions of life have to do with what is spiritual, or at least with what is invisible and immeasurable.

With what physical tools, for example, shall scientists ascertain the nature of the ultimate reality, whether it is spirit or matter? With what instruments shall they observe the origin of the cosmos? With what experiments shall they discover the meaning of life or the moral laws by which we should live? What kind of scope will permit them to scope out the afterlife or to behold the end of the universe? Now it is all too true that some scientists try to lend the prestige of science to their philosophical opinions, asserting, for example, that there is no god, or that man has no soul, or that the universe will one day become a lifeless dust-bin. But such affirmations are altogether unscientific, for the truth about these matters lies altogether beyond the reach of their disciplines, as indeed many honest scientists will frankly admit.

And so, because of its limited focus and methods we must conclude that natural science is not and cannot be the Teacher sent by god.

But what of *philosophy*? Surely in this time-honored discipline we have an excellent candidate for god's appointed Teacher. After all, what is philosophy supposed to do if not supply solid answers to the questions of life?

And yet it cannot. Such, in any case, was my own conclusion when, after devoting four years and thousands of dollars to the study of philosophy, I graduated from a major American university without a single conviction concerning any of the great questions of life. My alarm and dismay were exquisite.

Interestingly, not a few professional philosophers have reached the same melancholy conclusion.

Diogenes Laertius (ca. 300 AD) quotes Socrates as saying, "One thing only I know, and that is that I know nothing."

Montaigne agreed, asserting that "Philosophy is doubt."

Henri Bergson confessed, "Intelligence is characterized by a natural incomprehension of life."

R. D. Hitchcock concedes, "A modest confession of ignorance is the ripest and last attainment of philosophy."

John Seldon, adopting the same minimalist approach, opines that "Philosophy is nothing but discretion."

A story is told of pessimistic German philosopher Arthur Schopenhauer who, while visiting a greenhouse in Dresden, became so absorbed in contemplating a plant that his peculiar behavior elicited the concern of an attendant. "Who are you?" the attendant asked suspiciously. Schopenhauer replied, "Sir, if you could only answer that question for me, I'd be eternally grateful."

Similarly, someone once asked English philosopher Bertrand Russell if he would be willing to die for his beliefs. "Of course not," Russell replied. "After all, I may be wrong."[3]

All this would be funny if it weren't so sad. How is it possible that the one discipline charged with discovering answers to the questions of life can fail so completely in its mission? Are the postmodernists right after all? Is the greatest discovery of the "lovers of wisdom" that wisdom is not discoverable at all?

The test perspective, as we have already seen, supplies important answers to these urgent questions. It teaches us that man is indeed imbued with the philosophical spirit: sooner or later we all want to know the truth about the questions of life. But it also teaches us that the answers are not innate. They are not accessible by means of introspection, logic, or natural science. And this is just as true for philosophers as it is of the rest of us. All people—philosophers included—need a teacher sent by god.

The history of western philosophy only confirms these important conclusions. And yet, by surveying it for just a moment, we find that it does indeed supply a hint of a more fruitful road to travel.

Think of this history as a sandwich.

The bottom layer is the age of Greco-Roman philosophy (ca. 500 BC to 300 AD). It began when certain Greek philosophers cast off traditional mythological responses to the questions of life and sought to find answers through the use of unaided reason. Not surprisingly, as the years unfolded some of them turned to naturalism, others to pantheism, and still others to theism. In the end, however, they could not agree. Accordingly, as this period drew to a close, Greco-Roman philosophy was in a shambles, characterized by skepticism, cynicism, mysticism, and despair. The world was ripe for a new way of doing philosophy, a way that would not only revive the philosophical spirit but somehow satisfy it at last.

The middle layer of the sandwich is medieval Christian philosophy (ca. 300 AD to 1600 AD). During this era most people believed that a new way had indeed come. Philosophy thrived. Yes, there were differences of opinion as, for example, between traditional Catholics and various reformers. Nevertheless, all Christendom was united by a common philosophical culture. That culture was based on a common faith. All believed that God had revealed the answers to the questions of life by speaking to mankind through Christ and the Bible. For Christians, these two were God's appointed Teacher. Men may have disagreed about how to interpret the words of this teacher, but they did not disagree that the words had come from the one true god. Accordingly, this lengthy middle season in western philosophy was marked by creativity, contention, and even occasional confusion. But it was never marked by skepticism or despair. Because they had found a trusted spiritual teacher, philosophers—and the philosophical spirit—were alive and well.

The top layer of the sandwich is <u>modern philosophy (ca. 1600 AD to the present)</u>. For reasons we shall discuss later, this period began with a loss of confidence in the Bible and, indeed, with a rejection of the very possibility of divine revelation. The battle cry of the so-called Enlightenment was "Reason, not Revelation!" Men felt that in casting off revelation they were casting off superstitions that had trammeled the mind and hindered its search for truth. Like the Greeks and Romans of old, they were determined to turn away from the ancient Hebrew myths and turn instead to science, logic, and introspection. Here alone was the way to discover whatever answers we might need—including the answers to the questions of life.

Four hundred years of intellectual history now enable us to see what the *philosophes* of the Enlightenment could not—that their new way was actually an old way, and a counsel of despair as well. In taking the path of the Greeks and Romans, they arrived at the same destination as the Greeks and Romans. Just as before, some turned to naturalism, others to pantheism, and still others to speculative theism. In the end, however, they could not agree. And so, beginning in the 1950s, many philosophers finally gave up on the "modern" quest for truth—the quest for truth apart from divine revelation. Note carefully, however, that most of them did not turn back to revelation. Instead, they inaugurated the so-called *post*modern era, an era in which philosophy now courts its own destruction by abandoning the idea of truth itself. Some have hailed this as a great discovery. History shows, however, that it is simply the age of modern philosophy ending like the age of ancient philosophy—in a shambles characterized by skepticism, cynicism, mysticism, and despair. And among some, at least, it is also characterized by a desperate longing for a new and life-giving way of doing philosophy.

So again, in this briefest of surveys we find that the history of Western philosophy confirms what the test perspective teaches: the answers to the questions of life are not innate, so that all men need a divine revelation, a teacher sent from the unknown god. Accordingly, seekers cannot turn to philosophy—*or at least not to any philosophy that spurns divine revelation.* Rather, they must acknowledge the truth of G. K. Chesterton's words, that the mind is like a mouth: it is meant to bite down on something hard. That something is revelation. <u>Revelation is the philosopher's true food.</u> Just as the natural scientist was meant to feast on nature, so the philosopher was meant to feast on revelation. He can try to bite down on the world of nature, or on the contents of his own mind and emotion, but it will only hurt his teeth. What's more, if he continues to do so, he will starve. Here, then, is the philosopher's true wisdom: feast on revelation and live.

The Rough Road of Revelation

A seeker's journeys into all these spiritual *cul de sacs* can be deeply frustrating, but they need not be in vain. All that is necessary to make them profitable is for him to learn the lesson they teach: in his search for god's appointed Teacher, he cannot avoid traveling *the rough road of revelation.* However daunting, he must now begin to look for the person or group of persons through whom the unknown god may have been pleased to reveal his truth to the world.

Concerning this final phase of the search, there is both good news and bad.

The good news is that there is lots of revelation in the world. For example, we have much "theistic revelation"—revelation purportedly given by an infinite personal god. Included prominently in this category are Judaism, Christianity, and Islam, along with their many sectarian offshoots.

Then too there is "pantheistic revelation"—revelation supposedly coming from spiritually enlightened men, or possibly from disembodied spirits living on spiritual planes beyond our own. In this category we have the teachings of Hinduism, Taoism, and Buddhism, as well as revelations coming to us through various New Age "channelers" or spiritists.

The bad news, of course, is that these revelations are largely irreconcilable. In other words, they consistently offer conflicting answers to some, most, or all of the questions of life.

Now if logic counts for anything, this situation necessarily involves three possibilities: one of the revelations *probably* is true, some of them *must* be false, and (god forbid!) all of them *might* be false.

For mystics, however, logic doesn't count for much. It is, as Emerson famously said about consistency, the hobgoblin of little minds. Accordingly, the mystic sees another possibility: that world religions only *appear* to be contradicting each other, that they all are "really" saying the same thing, and that it therefore doesn't much matter which religion we practice, so long as we practice it sincerely. In short, since all roads lead to Rome, one road is pretty much as good as another.

If this viewpoint seems attractive, it is because there is an element of truth in it. All religions—to the extent that they acknowledge a spiritual ultimate reality—seek to understand and relate to that reality. They have glimpsed the unknown god and are attempting to establish a closer connection with him. But even if all religions share this common goal, it does not follow that all religions succeed equally well in achieving it. For example, one religion may tell us the true name of the unknown god (assuming he has a name), while another may tell us that he has no name or that he has many. One religion may describe him as he truly is, while another describes him as it thinks he is or as it wants him to be. One religion may enable seekers to establish a lasting connection with the (formerly) unknown god, while another may promise to do so yet continually leave them in shadow. In sum, one religion may actually be a dependable revelation in which a personal god reaches down to man, while another may be an undependable speculation in which man— peering through the semi-darkness of nature and conscience—falteringly reaches up to god. The result is that all religions may be one in aspiration while not being one in attainment.

Observe also that the mystic's understanding of religious diversity is always based on a pre-existing religious commitment, and that that commitment is nearly always pantheistic. How does the mystic "know" that all religions are really saying the same thing? It is because he "knows" that pantheism is true; that just as there is one Big Mind back of all (seemingly different) things, so there is one Big Mind back of all (seemingly different) religions. And why does the mystic smile condescendingly at seekers who carefully compare and contrast the teachings of different religions, hoping to find the one that is true? It is because he already "knows" that comparing and contrasting them is futile; that the discriminating intellect is actually an enemy; that common sense, reason, language, and even conscience all tend to divide reality into (the illusion of) multiplicity, whereas the true spirit of religion tends to dissolve all things into (the reality of) oneness.

The seeker, however, has made no such religious commitment and therefore "knows" nothing of the kind. In particular, he is not at all certain that pantheistic revelations are true. Accordingly, he cannot agree that all religions are "really" expressions of the one "perennial philosophy"—pantheism. Indeed, he finds it interesting and important that we must do great violence to the actual teachings of the theistic religions in order to pull pantheistic rabbits out of theistic hats. Reason, joined with careful study, persuades him that on nearly every question of life the theistic and pantheistic worldviews stand opposed; and from the test perspective he has learned to listen hard to the voice of reason. He knows it is important equipment from the unknown god, vital in his search for truth. How, then, can he follow the mystic by casting aside reason—and all the rest of his discriminating faculties—as useless obstacles in the pursuit of spiritual reality?

But if mysticism is not the way, what is? Again, it appears there is only one answer: seekers must confidently walk *the rough road of revelation*. Yes, human fallibility and duplicity have doubtless littered the spiritual landscape with religious refuse. And yes, it is even possible that evil spirits have contributed to the confusion as well—for nearly all world religions acknowledge the existence and deceptive activity of evil spirits. But none of this precludes the possibility that one of the religions is indeed god's truth, and none of it releases us from the obligation of finding out whether this is the case. Therefore, taking the rough road of revelation, the seeker must diligently sort through all the competing revelations until, god willing, he finds the one that is true. If he wants it badly enough, he will.

But how exactly is he to proceed in this search? What principles should guide him? And perhaps most importantly, where in the world should he begin?

By way of response, let me suggest four assumptions that a seeker won to the test perspective may reasonably make.

First, he may reasonably assume that the Teacher's identity will not clobber him over the head. From within the test perspective this assertion makes perfect sense, for if the unknown god made finding his Teacher too easy, the test would not be a test. Thus, a seeker should brace himself, understanding that significant effort is going to be required. He will have to dig deep—deep enough to uncover the god-given evidences by which he may at last be assured that this or that one is indeed the Teacher appointed by god.

Secondly, a seeker may also assume that the Teacher's identity will not be too obscure. This too makes sense. After all, the divine Tester is on our side. If he has sent us a teacher, it is because he wants us to find him. Yes, the Teacher may superficially resemble other teachers, just as wheat superficially resembles chaff or gold resembles pyrite. But in the end, anyone who really wants to find him can, even the simplest among us.

This assumption has practical ramifications. It means that god's appointed Teacher is likely to be a public person rather than a private, a herald rather than a hermit. It means that he will offer the kind of credentials that average people can respect; that he will use the kind of words average people can understand, and that he will make the kinds of demands with which average people can comply. In short, seekers may reasonably assume that god's appointed Teacher will not make himself available only to intellectual or spiritual giants, but to every honest soul, great or small, who is simply willing to keep his eyes open for truth.

Third, a seeker may reasonably assume that the unknown god will direct us to his Teacher by means of *supernatural signs*. For consider: if a divine Tester wanted to get our attention, how better than to use the unusual? And if he desired to draw us to his Teacher, how better than to surround him with the miraculous? Furthermore, if he desired to test our love of the truth, how better than to use phenomena which the lazy or recalcitrant could easily shrug off as fraud or superstition, but which the diligent and open-hearted, *after careful investigation*, could finally recognize as the handiwork of Heaven?

Seekers understand that the natural, moral, and probationary orders all point to an infinite personal god—precisely the kind of god who could use the supernatural to direct us to his Teacher. Believing this, they therefore have at their disposal an excellent way to begin their search: *they should keep their eye out for a teacher who is surrounded by supernatural signs.*

Finally, a seeker may reasonably assume that if the Teacher has *already* come into the world, he will be surrounded by a large number of spiritually satisfied disciples who have followed the signs to his feet. How could it be otherwise? For if indeed this is god's appointed Teacher, he will surely have brought to mankind all the truths and all the spiritual experiences for which the unknown god has prepared the human heart. And if seekers have truly found such things at this one's feet, why would they want to leave in search of another? They are seekers no more, but finders—finders who have come home. So then, those who have not yet come home do well to keep their eyes out for those who have.

Now before continuing to the next section, please take a moment to ask yourself the following important question: Who among all the world's religious teachers that you are familiar with best fulfills these criteria? Who, above all others, had a notably public ministry, connected well with the common man, was surrounded by supernatural signs, gained a large and committed following, and claimed to be bringing to the whole world god's answers to the questions of life? Think about it, write down your top two or three choices, and then read on to see how one of your fellow-seekers answered this question many years ago.

Window on a World of Signs

By and large, rumors deservedly have a bad reputation. Yet we must admit that rumors are often true and occasionally of great importance. Indeed, in a world such as ours the unknown god himself may not be above starting an occasional rumor if he thought it could help a poor seeker find his Teacher. He knows people talk. And he knows there is nothing like a few signs to get them talking—and moving toward the one whom he has sent.

So it happened with me back in the early 1970s. In those heady days of widespread spiritual inquiry I had become a seeker. I was deeply absorbed in the questions of life, especially the question of the ultimate reality. Through nature and conscience I had caught a glimpse of the unknown god. I had concluded that natural science and modern philosophy were indeed dead ends, that down those roads I would find no ultimate answers at all. And so my thoughts began to turn toward world religion. I realized that I must now journey down the rough road of revelation. But where in the world was I to begin?

As a matter fact, I began with what was then much "in the air"—Eastern religion. In my first year alone I studied and practiced Tibetan Buddhism, Hindu yoga and Zen Buddhism. But it was not long before something caught my eye, something supernatural. As I considered the teachings of my gurus and Zen masters, it began to dawn on me that there was one teacher who somehow stood out from all the rest: Jesus of Nazareth. Though I had taken little religious training as a child and no biblical instruction at all in college, I had heard enough rumors about Jesus to sense that he was unique. After all, had he not performed many astounding miracles? Had he not predicted the future? Had not the common people received him gladly? Was he not revered as the wisest of teachers and the best of men? And did he not have a large and enthusiastic following even to this very day, even in Santa Cruz, California?

And then there was the most amazing rumor of all—the story of his resurrection from the dead. Already I had read a great many "yogi books" about enlightened masters and god-intoxicated men. But never had I read or heard about any guru or teacher who had risen from the dead and then ascended bodily into the sky!

So, alerted by all these signs, I decided to go to where they pointed. I decided to learn more about Jesus and more about what he taught. This meant, of course, that I had to read the Bible. And so, for the first time in my life, I opened one up. When I did, I found to my amazement that I had actually opened a window—*a window on a world of signs.*

In a moment I will tell you more about the signs I saw. But first, for the sake of those who are unfamiliar with it, let me offer a few introductory words about the window itself, the Bible.

The Bible is actually a book of books, sixty-six of them. It was written by about forty different Jewish authors (plus Luke, a Gentile doctor), in three different languages, in seven different literary *genres*, over the course of some 1500 years (from about 1500 BC to about AD 60). Importantly, the stories it tells reference hundreds of different historical persons, places, things, and events. For these and other reasons, the Bible displays a very great diversity.

Yet it also displays an extraordinary—some would even say miraculous—unity.

For example, all of the books speak of *one god.* In the Old Testament (OT), he is called *Elohim*, the creator and sustainer of the universe. He is also called *Yahweh*, the covenant-keeping LORD of his people Israel. In the New Testament this same god is in view, but is further unveiled by Jesus and his apostles as a holy trinity, a three-in-one god eternally existing as Father, Son, and Holy Spirit.

Together, the books tell *one unfolding story*—a story of the *creation* of the universe, life, and man; their *fall* into evil, suffering, sickness, and death because of the sin of the first man, Adam; and their final *rescue and restoration* (i.e., *redemption*) by the triune creator turned redeemer. Needless to say, a story built around such themes should be of the greatest possible interest to seekers since it definitely touches on the questions of life!

Very importantly, the biblical books also affirm that the cosmic redemption is to be accomplished through *one central character*—the Messiah (or, in Greek, the Christ). This title means "The Anointed One." It is a term first used by the OT prophets to declare that God, in days ahead, would anoint the promised redeemer with his Spirit, thereby enabling him to accomplish his great work of saving the world, (Isaiah 42:1f, 61:1f).

As to his nature, the Bible teaches that the Messiah is both human and divine.[4] He is, in the picturesque language of the early Greek theologians, the *theanthropos*, the God-Man. More particularly, he is at once the human son of David (an ancient prototype of the *royal* Messiah) and the divine Son of God, (Matthew 22:41-46, Romans 1:1-6). This, the mystery of the Incarnation, is one of the great themes of NT theology. Over and again, the apostles marvel that God the Father has sent his divine Son into the world through the womb of a virgin, so that her human offspring, Jesus of Nazareth, might live, die, and rise again to redeem the cosmos, (Matthew 1:18-23, Luke 1, John 1:1-18, Philippians 2).

Concerning his work, the Bible portrays the Messiah as a cosmic redeemer who accomplishes his mission by occupying three offices familiar to Israelites of OT times: *prophet, priest* and *king*. As a prophet, he brings God's truth not only to Israel but to all nations, thus redeeming them from ignorance and error (Deut. 18:15-19, Isaiah 2:1-4, 9:2, 49:6). As a priest, he offers himself as an atoning sacrifice for the sins of his people, thus redeeming all who trust in him from divine condemnation, (Psalm 110, Isaiah 53, Zech. 6:12-13). And as king, he rules from Heaven in God's stead over the faithful of all nations, thus redeeming them from their sinful rebellion and autonomy, (Psalms 2, 110, Isaiah 9, Daniel 7:9-14). One day the king will descend from Heaven to redeem the material universe itself.

For the NT writers, the person and work of the Messiah are the central themes of all divine revelation. On this view, the primary characteristic of the so-called *Old Testament* books is that they look forward to the Messiah's coming. The primary characteristic of the *New Testament* books is that they celebrate his arrival in the person of Jesus of Nazareth, even as they continue to look forward to his return at the end of the age, when he will consummate God's redemptive plan by raising the dead, judging the world in righteousness, and renewing the entire cosmos.

And so, because of this amazing, multi-layered, Christ-centered unity, Christian interpreters see the sixty-six books as one book, *the* Book, the Bible.[5]

But there is more. For the Bible also discloses *one* (very large) *body of supernatural signs*—signs attesting that Jesus of Nazareth is indeed the promised Messiah. Here again is something to perk up the ears of every alert seeker. For if these signs are credible, would they not suggest that the unknown god and Israel's God are one and the same? Would they not identify Jesus as his appointed Teacher? And if, after all this, Jesus actually *claimed* to be the Messiah, would he not be identifying himself as the supreme prophet—which is to say, the god-appointed Teacher—of the entire human race?

All of this brings us back to the window and to what I saw when I first looked through it: I saw the one body of Messianic signs. Much indeed could be said about them. Here, however, I want simply to list them, adding no more than a few explanatory notes. In so doing, I hope to give you a feel for their abundance, diversity, supernatural-ness, and, most importantly, their amazing convergence in the person of Jesus of Nazareth. I also hope that they will impress you as much as they did me.

Note carefully that there are actually hundreds of signs, but that they readily fall into the following eight categories.

Signs Surrounding Jesus' Birth—These include angelic visitations, his birth to a virgin, a revival of the spirit of prophecy among certain devout Jews of Jesus' day, and the God-inspired journey of the Persian wise men to the place of his birth.[6]

Angelic Visitations and Testimony—These include angelic annunciations of Jesus' soon-coming birth, angelic ministry to Jesus in the wilder-ness and in the Garden of Gethsemane, and angelic appearances at his tomb and on the mountain near Jerusalem from which he ascended into the sky. Under this heading belong also the terrified confessions of demons—fallen angels with whom Jesus did battle in the days of his ministry to Israel.[7]

Theophanies—A theophany may be defined as a manifestation of God in which he sensibly displays his presence to men. During Je-sus' earthly ministry, God thus showed himself twice: once at Jesus' baptism and once again at his transfiguration on a high mountain in Galilee. The apostle Peter was present at Jesus' transfiguration when God manifested himself both visibly and audibly. Writing of this ex-perience toward the end of his life, Peter commended it to his fellow

Christians as one of the outstanding proofs that Jesus of Nazareth is indeed the God-authorized Teacher of the human race.[8]

4 _Miracles_—Miracles may be defined as extraordinary happenings that are designed by God to reveal his glory and help his people. According to the apostle John, Jesus performed so many miracles that all the books in the world could not contain them! These include healings, exorcisms, resuscitations of the dead, acts of power over nature, clairvoyance, and numerous predictions of the future. The New Testament authors speak of numerous miracles performed through the apostles, and also assume that at least some of Jesus' disciples will be empowered by Christ to perform them until his return at the end of the age.[9]

5 _The Resurrection_—Here is the most unusual and important of all biblical miracles—that Jesus died, was buried, and rose again from the dead after three days. His was not a mere resuscitation. The biblical authors record that after his resurrection he showed himself to hundreds of eyewitnesses over the space of forty days and then ascended visibly into Heaven. In other words, they affirm that Jesus rose to eternal life. Not surprisingly, his followers attached great importance to the resurrection, citing it often as the preeminent sign of God's favor upon their master. For them, it was the single most important reason why seekers of divine truth should turn, come, listen, and learn at Jesus' feet.[10]

6 _Old Testament Messianic Types_—The Old Testament consists of the thirty-nine biblical books written before Jesus' coming. The work of many different authors over the course of about 1100 years, these books contain numerous Messianic types. A Messianic type (Greek, _tupos: form_ or _figure_) is an OT person, place, object, event, or institution that symbolically points ahead to the Messiah and to the events of his life. Jesus and his disciples firmly believed that he was the promised Messiah, and that the OT types all found, or will yet find, their fulfillment in him.

A single example will give a feel for the nature of such types. In the OT book of Numbers we learn that the Israelites, recently escaped from Egypt, were grumbling against God in the wilderness (Numbers 21). As a result, God judged them by sending poisonous serpents into their camp. When the people cried out to God for mercy, God told Moses to make a bronze serpent and suspend it on a pole. Looking upon it, those stricken by the serpents would be healed. Jesus saw this entire episode as a type of himself and his work. Like the serpent, he too would be lifted up on a pole, bearing the sins of his people, so that all who look upon him in simple faith may experience forgiveness, spiritual healing, and eternal life, (John 3).

The NT authors find many such types in the OT, while Christian interpreters down through the centuries feel sure they have unearthed many more.[11]

Old Testament Messianic Prophecies—These are explicit OT predictions of the person and work of the coming Messiah. Again, both Jesus and his disciples affirmed that the Messianic prophecies have been, or will yet be, fulfilled in him. Christian interpreters argue that *the entire course of Jesus' life was foretold in OT prophecy:* his divine preexistence as the Son of God, his virgin birth, his birthplace, his miraculous ministry to the downtrodden, his death on the cross, his resurrection, his ascension, his reign in Heaven, and his coming again in power and glory at the end of the age. Note that OT types and prophecies point not only to Jesus but also to the divine inspiration of the biblical books in which they have been preserved.[12]

The Church—Jesus called his followers his Church, promising that they too would become signs. Their words, supernaturally transformed character, good works, growing numbers, and perseverance in the face of persecution and martyrdom would point seekers everywhere to the master that his followers knew and loved. The book of Acts records the birth and early history of Jesus' Church. Two thousand years of Church history record the rest. Here, then, is a unique sign, for it is seen not only through the window of the Bible but also down through history and around us in the world today.[13]

As I said earlier, in my first reading of the Bible (and especially of the four Gospels) I was deeply impressed by all these Messianic signs. I did not, however, fully appreciate their importance. Enjoying now the benefit of further reflection, let me conclude this section with three observations that should be of special interest to seekers.

First, seekers should understand that *the Christian faith is altogether unique in commending its truthfulness to the world by signs.* All religions claim to be true. Some even ascribe supernatural phenomena to their founders. But none—with the exception of Christianity—issues its truth claims, cites a wide variety of supernatural evidences in their support, and then explicitly challenges people to examine them. Interestingly, Jesus himself laid down this very pattern (John 5), while the New Testament reveals that his disciples faithfully followed it. So too have Christian preachers and teachers down through the centuries. In so doing, all have operated on the assumption that life is a test of our love of the truth, and that pondering and following the signs granted by Israel's God is one of the best ways to discover what that truth is.[14]

Second, *the biblical signs create a reasonable presumption that Jesus of Nazareth is indeed god's appointed Teacher.* This important conclusion follows logically from the very great abundance of signs, their amazing diversity, their having been spread out over some 6000 years of human history, their appearance in highly credible historical accounts, and, above all, their marvelous convergence in one man, Jesus of Nazareth. This phenomenon is unprecedented and unparalleled in all religious history. If, as Christian philosopher Os Guiness has written, Jesus is the world's greatest magnet for seekers, it is because those seekers are drawn by the world's greatest collection of signs.[15]

Finally, *Jesus confirms the implication of the signs by explicitly identifying himself as God's appointed Teacher.* Here, the last piece of the puzzle falls into place. The signs create a presumption that Jesus is the Teacher; Jesus confirms the presumption by declaring that indeed he is the Teacher.

His own words leave no doubt about his views on this crucial matter. Let us look at just a few of them.

First, Jesus saw himself as a *unique teacher*, "a greater than Solomon" who was sent into the world to unveil mysteries that the ancient prophets longed to see and hear, (Mt. 12:42, 13:16-17).

Next, he saw himself as a teacher *bringing God's full and final revelation*, telling his disciples that he had given them God's own words (John 17:6-8), and that in days ahead he would teach them "all things," (John 15:15, 14:26). In this confidence, he also proclaimed that his words were henceforth the one rock upon which alone men may safely build their lives, (Mt. 7:24-29, 28:18f). Indeed, men of every generation must keep building on them, even to the end of the age, when at last those same words will become the judge of all, (John 12:48).

Very importantly, Jesus saw himself as bringing this revelation not only to Israel, but *to all mankind*. Thus, echoing the ancient prophecy of Isaiah, he declared to the citizens of Jerusalem, "I am the light of the world," (John 8:12, Isaiah 49:6). On another occasion, when a Samaritan woman said to Jesus, "I know that the Messiah is coming… he will declare all things to us," Jesus replied, "I who speak to you am He," (John 4:25-26).

Similarly, when Pilate interrogated him about his claims to royalty, Jesus answered, "You say correctly that I am a king. For this cause I was born, and for this cause I have come into the world, that I should bear witness to the truth. Everyone who is of the truth hears my voice," (John 18:36-8). Here Jesus sought to bring God's truth to a Gentile—a Gentile who refused to hear his voice, (John 18:38).

Finally, and quite mysteriously, Jesus saw himself as God's appointed Teacher *even to the end of the age*. This is reflected in his final words to his disciples. For example, Jesus warned them, "Do not be called Rabbi, *for One is your Teacher*, and you are all brothers… And do not be called leaders, for One is your Leader, *even Christ*," (Matthew 23:8-12). Here we see that he identifies himself as the Christ, and that as such he intends to be the Teacher of his people right up to the time of his return, (Matthew 28:20). Elsewhere in the NT we learn how he plans to accomplish this spiritual feat: after his return to Heaven, he will send the Holy Spirit, through whom he (Jesus) will enable his apostles to complete the written revelation of God's truth. After that he will send the Spirit to all the rest of his disciples, so that they in turn may be able to understand it, (John 14:15-18, 25-26, 16:12-13, 15; see Luke 24:45).

Summing up, then, we find that Jesus of Nazareth saw himself as a unique teacher sent from above to bring a full and final revelation of God's truth to all mankind—a revelation that he himself would enable his apostles to receive and write down, and the rest of his disciples to recognize, preserve, understand, obey, and delight in, even to the end of the age.

A testimony like this, confirmed by so great a wealth of supernatural signs, is a bright and shining star in the sky of world religion. It can hardly fail to turn a seeker's compass toward Nazareth.

At the Feet of the Teacher

On the strength of the signs, it is certainly reasonable for a seeker to begin his search for divine truth with Jesus of Nazareth. And on the strength of his testimony about himself, it is all the more reasonable to assume that Jesus is—or is very likely—god's appointed Teacher.

Yet despite all the evidence, one vital step remains. For though all the world were filled with signs, a seeker could never be fully satisfied until he knew that Jesus had answered all or most of the questions of life, and that he had answered them well. Thus, the seeker's next logical step is to come, sit at Jesus' feet, and hear and evaluate what he has to say.

Concerning this climactic stage of the journey, several important points may be made.

First, it is evident that any evaluation of Jesus' teaching must be undertaken with great thoughtfulness and deep humility. The reasons are many. We have already seen, for example, that mankind apprehends spiritual truth with considerable difficulty and that our thinking is vulnerable to error and bias. Thus, in the case of a supernaturally attested teacher like Jesus, the path of wisdom is surely to doubt one's own views before doubting his. This is all the more true when we step back and look at things in historical perspective; when we see how mankind's philosophical and scientific opinions are "in" one day and "out" the next, while in every generation Jesus' teachings continue to find skilled defenders and a large, devoted following. In short, many factors warn the wise against a rush to judgment.

On the other hand, despite its perils we must also understand that critical investigation is absolutely necessary. For again, the seeker's ultimate goal is to find the truth and to see that it is true. How, then, can he avoid hearing and evaluating what purported truth-tellers have to say? On the premise that life is a test, the perils of judging god's appointed Teacher cannot possibly preclude an honest evaluation of his teaching since the test perspective positively demands it.

The necessity of such evaluation contributes richly to the drama of being human. Yes, our poor faculties may be wounded, perhaps far worse than we realize. Yet for all that we must go forward, trusting that he who created those faculties will help us to use them effectively, if only we will do so humbly, honestly, and persistently. In short, Jesus may indeed be god's appointed Teacher and any human evaluation of his views a kind of folly. But if the divine Tester has ordained careful investigation as part of the test of life, then it is a sublime folly and ready to receive its just reward.

But how, precisely, is a seeker to go about evaluating Jesus' teachings—or the teachings of anyone who presumes to bring us a revelation from god?

Here again we are much helped by the test perspective. For if life really is a test of our love of the truth, then we may be sure that the unknown god has adequately equipped us to weigh the truth claims of different teachers and religions. Indeed, from within the test perspective we suddenly begin to see the wonderful workings of our minds as the means to this very end. We see ourselves as having been fitted with "truth monitors"—faculties designed by the divine Tester for the express purpose of pondering and evaluating different answers to the questions of life. Ours is simply to use those faculties as best we can; his is to enable us to see the truth as we do.

But what exactly are these truth monitors? In *The Test*, I isolate four: intuition, reason, conscience, and the human inclination to hope for the best. Accordingly, I argue that a true revelation about cosmic origins, or any other question of life, must be:

1. *Intuitive*—It must not offend but rather win the assent of intuition; it must speak acceptably to what might be called our "spiritual common sense."

2. *Reasonable*—This criterion is actually three-fold. It means that a revelation must be understandable, that it cannot contradict itself (i.e., it must be logical, obeying the laws of sound thought), and that it must be supported with an abundance of good evidence. All this does not, of course, rule out mystery, in the sense of truth hidden from our sight or from our complete understanding. It does, however, rule out mysticism—revelations that offend or disparage the god-given faculty for apprehending and discussing truth: reason.

3. *Right*—It must not violate our conscience, but rather must commend itself to our distinctly ethical intuitions as consistent with the good and holy god who created and sustains the objective moral order.

4. *Hopeful*—It must awaken hope, not only the hope of finding true answers to the questions of life but also of laying to rest the spiritual anxieties and longings associated with each one of them. In other words, a true revelation from the unknown god must offer us peace of mind, both for this life and for the next.

And with this we come, at long last, to the book you now hold in your hands. For *In Search of the Beginning* is, as I said earlier, a much-expanded chapter taken from that part of *The Test* in which I introduce my readers to Jesus' teachings on the nine questions of life and then evaluate them in light of these four criteria. Needless to say, it's a big job. Moreover, in working at that job, I soon found that one part was bigger than all the rest: the part that concerns the beginning—the origin of the universe, life, and man.

The difficulty here stems largely from the historical situation in which we now find ourselves. There was a time not long ago when, throughout much of the world, Jesus' view of the beginning was taken for granted. That time is past. Today—and for approximately the last

150 years—the dominant view of origins among the western intelligentsia has been what I will hereafter call *cosmic evolution.* Indeed, many of us (myself included) were brought up hearing no other view, simply taking it for granted that this was the truth. If, therefore, it is *not* the truth, I feel quite certain that millions of modern seekers will require a good deal of persuading to become convinced that such is the case.

Here, then, is my main reason for turning a single chapter into an entire book: I have come to believe that Jesus' view of the beginning is very much worth a respectful hearing, yet I also know from painful personal experience how difficult it is for him to get it. Modern seekers will never be shaken from their evolutionary slumbers apart from an outstandingly clear, thorough, and persuasive defense of the biblical cosmology. A few have already been written. My goal in this book is to present another.

In our ramble through the lowlands, I have already labored much toward this end, setting out a concise case for the existence of an unknown god and for Jesus of Nazareth as his appointed Teacher. Nevertheless, several important steps remain. First, I must introduce my readers to Jesus' teaching on the beginning. Next, I must evaluate that teaching in light of the four criteria for a trustworthy revelation. And finally, I must also introduce and evaluate several other popular cosmological options. For it is only by placing all the legitimate contenders in the same ring that seekers will be able to see which one, if any, has the strength to prevail.

My approach to all this will be as follows.

In Chapter 1, I set the stage for our journey, drawing from personal experience to put my readers in touch with their own innate desire to behold the beginning, and to awaken in them a faith that they really can.

In Chapter 2, I critically examine naturalistic views of the beginning, focusing on the modern theory of cosmic evolution. Naturalists do not, of course, present their cosmology as a divine revelation, since naturalism by definition is an atheistic worldview. Nevertheless, they do present it as truth, or at least as the most reasonable approximation of truth we now have. And perhaps, despite their atheism, there is some truth in what they say. Therefore, even if a seeker disagrees with a naturalist's atheism, he still will want to examine his claims in order to see just how trustworthy they are. This is all the more necessary in view of the tremendous influence naturalistic cosmology has had upon the modern world.

In Chapter 3, I go on to examine two views of special interest to spiritually minded people—the cosmologies of classical Hinduism and the modern New Age movement. These pantheistic versions of the beginning are still much in the air. The latter in particular has attracted a large following, since it seems to invest the widely assumed cosmic evolution with spiritual significance and hope. Careful seekers will want to know, however, if these pantheistic cosmologies really meet the high standards of a trustworthy revelation.

In Chapter 4, we reach the climax of our journey, examining in some detail Jesus' views on the Genesis creation account; the identity of the creator; the origin, structure and purpose of the cosmos; its spiritual as well as physical components; and the vexed but crucial question of its age.

In chapter 5, I offer some thoughts about spiritual factors that contribute to the controversy surrounding cosmology today. Then, reflecting on some of my own failures as a seeker, I conclude by suggesting a five-fold way by which we can make a satisfying personal journey to the beginning, where, at long last, we may see it—and all it involves—for ourselves.

It remains only to add that I have included in the pages ahead a number of anecdotes taken from my own spiritual journey. These are largely drawn from a four-year period of intense spiritual searching that began immediately after my graduation as a philosophy major from the University of California at Santa Cruz, (1970-1974). During a portion of that time I studied Christianity under the tutelage of a Franciscan priest, Father Gabriel Barry. During most of it, however, I assiduously believed and practiced various Eastern religions, especially Zen Buddhism.

Reading the present book, you will no doubt guess how my journey ended. To know the whole story, however, you must await the publication of *The Test*.

NOTES

✳ 1. Throughout this book I use the word "god" when referring generically to the supreme being, the object of mankind's inquiries and speculations about an ultimate spiritual reality. On the other hand, I use the word "God" when referring to the god of the Bible. In so doing, I am using the word as the Bible does (and as we in the West have traditionally done)—as a proper name, the English equivalent of the Hebrew *Elohim* and of the Greek *Theos*. Thus, God is the god of the Hebrew-Christian Scriptures. Though a bit irksome at first, this distinction will prove quite helpful in the pages ahead.

2. C. S. Lewis, in the following excerpt, concludes from the experience of natural hunger that our spiritual hunger for something like Heaven is a good sign that Heaven exists. So too with our hunger for truth.

> Creatures are not born with desires unless satisfaction for those desires exists. A baby feels hunger; well, there is such a thing as food. A duckling wants to swim; well, there is such a thing as water. Men feel sexual desire; well, there is such a thing as sex. If I find in myself a desire that no experience in this world can satisfy, the most probable explanation is that I was made for another world.

See C.S. Lewis, *Mere Christianity* (Harper San Francisco, 2001).

3. These quotes were included in an article by Dr. George Fox, entitled "The Philosopher's Dilemma!" It appeared in *The Grace Messenger Newsletter*, (Fall, 2000). Contact Grace School of Theology, 40 Cleveland Road, Pleasant Hill, CA 94523.

4. Christian interpreters find both hints and explicit affirmations of the divine nature of the Messiah throughout the OT. See Psalms 2, 110, Isaiah 7:14, 9:6-7, Jeremiah 23:5; Daniel 7:9-14; Micah 5:2; Malachi 3:1.

NT passages affirming or implying the deity of Jesus of Nazareth include Matthew 1:23, 11:25ff, 22:41-46, 24:30-31, 28:20; Mark 2:1-12; John 1, 5:16-33, 6:44, 8:46, 8:58, 9:35-36, 16:30, 15:5, 20:28; Philippians 2, Colossians 1, Hebrews 1-2, Revelation 1-3.

The NT doctrine of the trinity is seen vividly in Matthew 3:13-17, 28:18ff, John 14:15-19, 23-24, 16:13-15, 17:20-21, 2 Corinthians 13:14, 1 Peter 1:1-2.

5. For many Christians, the Christ-centered unity of the Bible is the supreme proof of its divine origin and trustworthiness. How, they ask, could some forty different authors, writing over the course of 1500 years, manage to create a single story, about a single god, administering a single plan of salvation, through a single redeemer (the Messiah), who is attested by a single set of supernatural signs, and who is worshiped by a single people, according to a single (and eminently satisfying) worldview? Such intricate, multi-layered unity seems to permit but one answer: a single divine Author must have superintended not only the *creation* but also the *preservation, recognition,* and *final collection* of the 66 books that we now call The Book, the Bible.

The unity of the Bible, so compellingly supernatural, also supplies a basis for much that Christians believe about the character of their Book. They say, for example, that the Bible's unity entails its divine *inspiration*—for how, apart from such inspiration, could its several authors have produced its many-faceted oneness, (2 Timothy 3:16-17)? But if the Bible is inspired, then it must also be *inerrant* in all it affirms—for how could a divinely inspired book be in error, (John 17:17)? And if the Bible is inerrant, then it must also be *complete*—for both Christ and his apostles (inerrantly) taught that through themselves, and themselves alone, God was at last bringing to all nations *a full and final revelation* of his spiritual truth, (Matthew 28:18f; Ephesians 2:19-20; Jude 1:3; Revelation 22:18-19). It is, then, because of the Bible's astonishing unity that many Christians embrace it as the very Word of God. (See Appendix 1).

See Dean Davis, "One Shot, One Book, One God," *Journal of the Christian Research Institute,* (December, 2004).

6. See Isaiah 7:10-14; Matthew 1:18-25, Luke 1 and 2; Matthew 2:1-18, 4:3, Mark 3:11.

7. See Luke 1, 2, 22; Matthew 28, Acts 1.

8. See Matthew 3; John 1; Matthew 17, Luke 9, 2 Peter 1.

9. See John 21:25, Acts 2:22, 3:1-10; 1 Corinthians 12:7-11.

10. See Psalm 16:10; Isaiah 53:11; Hosea 6:2; Mark 9:31, 10:34, John 2:18-25; Matthew 28; Mark 16; Luke 24; John 20, 21; 1 Corinthians 15:1-11.

11. See appendix #2.

12. Isaiah 7, 11, 48:16, 49:1-6, 53, 61, 63; Micah 5; Isaiah 61; Isaiah 53; Psalm 2, 16, 110; Genesis 3:15.

13. Acts 1:7-8.

14. We have a striking example of this pattern in Peter's sermon on the day of Pentecost, (Acts 2:21f). Seeking to win his Jewish brothers to faith in Jesus, Peter declared:

> "Men of Israel, listen to these words: Jesus of Nazareth, a man attested to you by God with miracles and wonders and signs which God performed through him in your midst… this man you nailed to a cross by the hands of godless men and put to death. But God raised him up again."

Here, Peter commends the truth of Christianity to the Jews on the basis of Jesus' miracles and resurrection, and then goes on to cite several OT predictions of those very events. In all this, Peter was simply following in the footsteps of his master, who had himself cited both his miracles and the OT scriptures as proof that the Father had sent him into the world as its authorized prophet (teacher), priest, and king, (see John 5:31f; Acts 10:34-33, 17:22).

15. Many biblical scholars reject the virgin birth, angelic visitations, theophanies, miracles, and the resurrection of Jesus as myths and legends. They do so, however, not because solid historical evidence leads them to that conclusion but because they themselves do not believe that such things are possible. One among them, Rudolph Bultmann, declared flatly, "The continuum of historical happenings

cannot be rent by the interference of supernatural, transcendent powers. Therefore, there is no 'miracle' in this sense of the word." Seekers, however, find such dogmatic skepticism impossible to embrace. They have glimpsed a personal god behind the natural, moral, and probationary orders. If such a god exists, why should he not be able to act supernaturally if he so desires? Indeed, it is only reasonable to expect that he *will* act supernaturally—for if life is a test, then supernatural signs are just the thing to lead us to the god-authorized Teacher who can help us pass.

Seekers must, of course, be duly cautious in evaluating the world's miracle stories, for some may indeed be legendary, while others may be historical in nature but demonic in origin. Still, seekers cannot rule out miraculous signs altogether. Rather, on a case-by-case basis, they must try to determine if there is credible historical evidence to support the sign, and also if this sign nourishes hope and stimulates godly living. Such evaluations may seem difficult, but if the unknown god has in fact granted true signs, then seekers can trust him also to grant sufficient outward evidence and inward illumination for us confidently to distinguish the true from the false. The only condition is that we want to.

CHAPTER 1

In Search of the Beginning

It is, I now firmly believe, the birthright of every child to behold the beginning: the origin of the universe, life, and man. But as my own story makes painfully clear, in today's world receiving the vision can be the work of a lifetime.

Like all children, I was fascinated with the beginning. I wanted to go back—all the way back—to that special place where I could see it for myself. I knew that such a place existed, that a journey to it was possible, and that the scenes awaiting me at its end were glorious. It was only a question of finding the right road. And with Dad, Mom, Pastor Edwards, and my trusted school teachers to point the way, there could be no problem at all. They would lead, I would follow. The future of my journey into the past looked exceedingly bright.

My first road passed by way of Sunday school. Though I attended only sporadically, I was nevertheless quite clear about what my Bible Story Book had to say about the beginning. To this day I can still see the picture of Adam and Eve. Situated amidst garden verdure, surrounded by tame animals, youthful, clear-eyed, smiling, healthy, fair, and completely oblivious to a nakedness discreetly concealed by draping hair and strategically placed fronds—yes, the beautiful couple really were God's crowning touch, the capstone of His six days of creation. Moreover, they were clearly the center of His creation. Like planets orbiting the sun, all things somehow seemed to revolve around them. To my youthful mind, the picture made perfect sense. What's more, my heart received it gladly, for it was altogether bathed in light, order, clarity, simplicity, beauty, and (best of all) complete well-being.

It was not, however, affirmed by the world around me. I soon discovered, for example, that my parents did not believe it. Nor did my elementary school teachers. Indeed, I doubt today that even my Sunday school teachers believed it. All the authorities in my life seemed to admire the biblical beginning; none received it as the truth.

Now no matter how lovely the road, a child will not walk it alone for long. And so, in due season, the biblical story became just that—a story. Meanwhile, I found myself walking with the rest of the world down a very different path. The truth about the beginning, it appeared, was to be found at the end of a road called "evolution."

That journey began with a two-volume set of big red books given to our family by my erudite Aunt Ethel. Entitled *A Child's History of the World*, the glistening tomes seemed to promise all that my budding young intellect longed for—a comprehensive look at "the big picture" of history from the beginning right up to the present. But as I opened to the first chapter (*How Things Started*) my zeal for universal history soon received a near-fatal blow.

Warming to his task, the author invited me to go back with him to a time so long ago that "… you might say 'long, long, long' all day and all next week and all next year and it would not be enough." Yet despite this promising invitation, he never really did take me to the beginning. Instead, we regressed to a time when there was no world at all—only the stars, one of which was our sun.

So, with the sun standing in for the beginning, he now traced the origin of the world, life, and man.

The sun threw off a spark. The spark cooled and became our earth, a bare rock swathed in steam. The steam turned to rain and the rain produced the oceans. Then, in the oceans, "came" the first living things, tiny plants. The plants eventually migrated to the rocks. Then came the first tiny animals, "wee mites, like drops of jelly." Then came insects. Then came fish, and after that, frogs. Next came snakes and huge lizards, "bigger than alligators, more like dragons." Then came birds. Next came the mammals, "… like foxes and elephants and cows that nurse their babies when they are born." When the monkeys came, the stage was set for the grand finale:

> And then, last of all, came—what do you suppose?—yes, PEOPLE—men, women and children! So here are the steps, see if you can take them: Stars, Sun; Sun, Spark; Spark, World; World, Steam; Steam, Rain; Rain, Oceans; Oceans, Plants; Plants, Mites; Mites, Insects; Insects, Fish; Fish, Frogs; Frogs, Snakes; Snakes, Birds; Birds, Animals; Animals, Monkeys; Monkeys, People. AND HERE WE ARE!"

Already I was confused. Where had the stars and the sun come from? Just how did one thing "come" from another? Who or what was causing all this "coming" to come to pass? The red book seemed to raise more questions than it answered.

Finally, I arrived at the description of "early man." And with this, my fervent hopes for a comforting vision of the beginning were dashed once for all.

> These first Stone Age people were primitive. Primitive people were wild animals. Unlike other wild animals, however, they walked on their hind legs. They had hair growing, not just on their heads, but all over their bodies, like some shaggy dogs. They had no house of any sort. They simply lay down on the ground when night came. Later, they found caves where they could get away from the cold and storms and other wild animals. They spent their days hunting some animals and running and hiding from others. They lived on berries and nuts and grass-seeds. They robbed the nests of birds for eggs, which they ate raw, for they had no fire to cook with. They were blood-thirsty; they liked to drink the warm blood of the animals they killed, as you would drink a glass of milk.

> They talked to each other by some sort of grunts—"Umfa, umfa, glug, glug." They made clothes of skins of animals they killed. Though they were real men, they lived so much like wild animals we call such people savages. They were fearful and cruel creatures, who beat and killed and robbed whenever they had a chance. A cave man got his wife by steal-ing a girl away from her own cave home, knocking her senseless, and dragging her off by her hair, if necessary. The men were fighters but not brave… their only rule of life was hurt and kill what you can, and run

from what you can't. This is what we call the first law of nature—every man for himself.

Suppose you had been a boy or girl in the Stone Age, with a name like Itchy-Scratchy... Your cave would have been cold and damp and dark, with only the bare ground or a pile of leaves for a bed. There would probably have been bats and big spiders sharing the cave with you... You might have had on the skin of some animal your father had killed, but as this covered only part of your body, and there was no fire, you would have felt cold in winter, and when it got very cold you might have frozen to death... There was nothing to do all day long but watch out for wild animals—bears and tigers—for there was no door with lock and key, and a tiger, if he found you out, could go wherever you went and "get you," even in your cave.

And then some day your father, who had left the cave in the morning to go hunting, would not return, and you would know he had been torn to pieces by some wild beast, and you would wonder how long before your turn would come next.

Do you think you would like to have lived then?[1]

Yes, here was my first encounter with naturalistic evolution and its widely accepted version of the beginning. To my tender sensibilities it seemed the exact opposite of the Bible's: confusing, meaningless, repulsive, and terrifying. This was particularly true of my brutal and short-lived ancestors. The more I contemplated it all, the more the distant past seemed like one of the caves in which they supposedly lived: dark, dangerous, and redolent of death. I recoiled. Yes, somewhere deep down I still longed to see the beginning. But according to the big red book, that meant going back into a cave. No, it meant going back into a tomb. And as a comfort-oriented ten-year-old, I had little interest in returning to a tomb.

So I gave up my search for the beginning.

The surrender was not, of course, instantaneous. But it was inevitable. Every authority figure in my world reinforced this picture. Had not my learned aunt and trusted parents introduced me to it? Did I not see it in my textbooks at school? Was it not found in encyclopedias, newspapers, magazines, and in nearly every issue of the ubiquitous *National Geographic*? Why, even Walt Disney agreed (*Fantasia*), to say nothing of *The Flintstones*! Yes, the red book must be right. But if my origins really were to be found in evolution's cave, that did not mean I had to visit often. Others could go back if they wished. As for me, I would turn to the light. I would live in the present and look to the future. Two-thirds of time would have to suffice.

This flight continued throughout my junior and senior high school years, even as the evolutionary picture of the beginning took stronger and stronger hold on my imagination. Now I was presented with the accepted scientific evidences for evolution: fossil remains in the geological column; human embryos "recapitulating" the stages of mankind's evolutionary history; anatomical similarities between animals; experiments in plant and animal breeding; vestigial organs—those useless remnants of our evolutionary past (e.g., tonsils, the appendix, wisdom teeth, etc.); and, most disturbingly, the skeletal remains of various "hominids," the ape-like ancestors of man.

Had anyone around me challenged the reliability of these evidences (e.g., the recapitulation theory or the uselessness of vestigial organs), or suggested other ways of interpreting them (e.g., anatomical similarities based on common design rather than common ancestry), I might have realized that evolution was not a proven fact at all. But the "icons of evolution"—the verbal and artistic reconstructions of the proofs and events of evolutionary history—had done their work. These pictures, encountered over and again throughout my childhood, became the lens through which I would henceforth view and interpret the actual facts of nature. No one I knew offered another interpretation or another set of pictures. I therefore graduated from high school a convinced, if unenthusiastic, evolutionist.

The situation did not change in college. Now one might expect that at a major American university the evolutionary paradigm—so far-reaching in its implications for every department of human inquiry—would be subjected to the most careful scrutiny. Yet my experience was precisely the opposite. On my campus, like most in the nation, the evolutionary beginning had attained the status of a presupposition, an axiom, a self-evident proposition—so much so that no self-respecting intellectual would dare to question it. The once reigning biblical paradigm of divine creation, if mentioned at all, was dismissively cited as a relic of the ancient, unenlightened past. It was understood to represent a primitive, mythological stage in mankind's (evolving) thinking about origins. Having no desire to associate myself with outdated myths, I remained an evolutionist.

Nor did I abandon evolutionism during my post-graduate years, when I attached myself to various pantheistic religions. I did, however, experience a fundamental change in the way I looked at it. By moving from a naturalistic to a pantheistic worldview, cosmic evolution suddenly took on a new and exciting meaning. Now there was nothing "random" or purposeless about it at all. Now, after so many eons of time, the hidden plan of Big Mind (the god of pantheism) was coming to light. Now, in man—who is the cutting edge of the evolutionary thrust—Big Mind was finally becoming conscious of itself as god!

This sudden conversion to evolutionary pantheism did not, of course, alter my basic understanding of the beginning. In those days, Big Bang cosmology was much in the air and I accepted it without question. But had I ever paused to give my new cosmology some thought, I would soon have realized that reconciling Big Mind with a Big Bang requires a Big Leap of faith. Indeed, as we shall see later, such a reconciliation is impossible even to conceive, let alone demonstrate.

But so great was my excitement that I did not pause—either to consider the metaphysical and ethical problems of evolutionary pantheism, or the slim and controversial scientific evidences supporting the Big Bang, or the troubled history of modern cosmology, or the increasingly vigorous debate as to whether cosmic evolution had occurred at all. No, I simply trusted my pantheist friends and teachers. They assured me that the marriage of Eastern religion and Western evolutionary cosmology was a good one, a match made in Heaven. Besides, why get entangled in the arcana of science and philosophy when all such reasonings are actually obstacles to the supreme goal—a mystical experience of the ultimate reality that is completely beyond the reaches of word, image, and human thought itself?

So once again—this time as a newborn pantheist—I embraced cosmic evolution. Now, however, I gladly saw myself as part of it. True, I still could not find my way back to the beginning. But no matter.

The future was beckoning, and that was far better. My friends and I were sons of the Age of Aquarius. Now was the appointed season in cosmic history when tired old humanity, led by visionary youth such as ourselves, would make an evolutionary leap into divinity itself.

But what, I anxiously wondered as I began my catechism in 1971, *would Father Barry have to say about THAT?!*

Lessons to Learn

My own turbulent history of seeking out, wistfully abandoning, fleeing from, reconciling myself to, and altogether ignoring various versions of the beginning has, of course, been played out in the lives of multitudes of modern men and women. What can we learn from this strange scenario? And how can the test perspective help us to understand it?

To begin with, we learn that man is perennially fascinated by the beginning. We humans have a mysterious capacity for time travel, as well as a powerful desire to go as far back in time as we possibly can. This desire first manifests itself when a child asks, "Daddy, Mommy, where did I come from?" or, "Where did people come from?" or, "Where did the world come from?" or, (if Daddy and Mommy are theistically inclined), "Where did God come from?" Happy are the parents who can confidently answer questions such as these!

The same desire is at work in adults. One person is strangely moved to go on a pilgrimage to the land of his ancestors, another to find her birth parents, another to flesh out his family tree by in-depth genealogical study. Observe also the scholars—historians, anthropologists, paleontologists, biologists, geologists, astronomers, and cosmologists—all of whom try to use present phenomena as a kind of telescope by which they might peer through the mists of time into the distant past. I call them adults, but in the end perhaps they too are children, children like Hansel and Gretel: secretly feeling themselves to be lost in the forest deeps of the cosmos they look urgently for any crumb of evidence by which they might make their way home to the beginning and to the father/creator who awaits them there.

Second, through a close examination of the workings of our own mind, we learn that the search for the beginning is propelled by a richly significant connection between this and other vital questions of life. For example, if we could somehow get back to the beginning, surely we would discover the nature of the ultimate reality—whether it is eternal matter, Big Mind, or an infinite personal creator god. Likewise, we would discover the true nature of "penultimate reality" (i.e., everything that comes from the ultimate reality)—whether it is simply the physical universe that we experience with our senses, or something more. And we would also understand the relationship between the two—whether penultimate reality arose by way of a transformation of eternal matter, an emanation of Big Mind, or a divine creation "out of nothing."

Beyond all this, we might also be able to learn something about the purpose of the universe (if there is a purpose), or the way man was intended to live (if there is an intention), or the destiny of the universe (if there is a destiny). Yes, the beginning is clearly a place rich with meaning and (we somehow sense) big with blessing. Therefore, with not a little existential urgency, we yearn to make our way back to it.

As elsewhere, we find here that the test perspective is both helpful and encouraging. For if indeed an unknown god has planted within us a desire to behold the origin of things, then no doubt he means to satisfy it. Can we not, therefore, count on him to grant us some kind of revelation enabling us to make the journey home? In other words, is not the desire to behold the beginning an implied promise that we really can—once we have sought out and finally found the trustworthy revelation by which the unknown god has been pleased to show it to us?

Here, by the way, is why I now believe that a vision of the beginning is the birthright of every child: because the test perspective implies that it is. And if the test perspective is true, then dads (such as myself), moms, teachers, clergy, scientists, publishers, journalists, filmmakers, and anyone else who presumes to educate children, all need to ask themselves some probing questions. Here is a sampling, guaranteed to make many of us uncomfortable:

Before teaching children about the beginning, do we not have a moral obligation to find out the truth of the matter for ourselves?

If we have not yet found it out, should we not hold our peace till we do?

If we feel compelled to present some cosmological options, should we not at least refrain from representing our educated guesses about the unobservable past as the hard facts of science?

If there really is a divine Tester, would it not be wrong—and possibly quite injurious—to mislead his children about their origins? Indeed, would it not be a species of blasphemy to deposit in the place reserved for divine revelation the transient, erroneous, and confusing conjectures of mere men?

And finally, what might be the attitude of the Tester toward those who do?

But alas, for all their weightiness, considerations like these are not likely to temper the declarations of most modern educators. And if life really is a test, that, somehow, is as it must be. Speculative scientific cosmologies will continue to come and go. New "revelations" will continue to pour forth. Even to the end of the world, children will meet and be confused by different versions of the beginning. The unknown god, it would appear, has wisely ordained it so. Accordingly, children of all ages must settle it in their hearts that they are going to have to search—and persevere in searching—till they actually find what they are longing to see.

Thankfully, in this endeavor there is every reason to be hopeful. Yes, the search for god's appointed Teacher may be difficult. And yes, receiving his revelation about the beginning may be more difficult still, challenging presuppositions deeply ingrained from childhood and widely established in the surrounding culture. Indeed, receiving his revelation might even threaten to marginalize a seeker, exposing him to ridicule, ostracism, and financial and vocational duress.

But if he truly is a seeker, all this would be as nothing compared to the promise held forth by the test perspective: a promise of finding the Teacher, falling in at his side, and walking with him all the way back to the beginning; a promise of his lifting the shroud that has somehow settled over the cosmic past, and of viewing the primordial scenes for which the creator himself has prepared our childlike hearts; a promise of being able to make this journey over and again, with gratitude and joy, to the very end of one's life; and (in some ways best of all) a promise of being able to help other children, both young and old, make the same life-giving journey for themselves.

In the pages ahead we shall see that Jesus of Nazareth invites us on just such a journey. Hopefully, all who learn of his invitation will judge the offer well worth any of the costs involved.

NOTES

1. V. M. Hillyer, *A Child's History of the World*, (Spencer Press, 1951), pp. 3-16.

CHAPTER 2

Naturalism on the Beginning

In seeking for truth about the beginning, we venture onto the terrain of cosmology. This discipline may be broadly defined as the study of the origin, structure, purpose, and destiny of the cosmos. In the chapters ahead I focus largely on what is called _cosmogony_, the study of the origin and structure of the universe, life, and man. The remaining aspects of cosmology I will address in various chapters of *The Test*.

Let us begin by surveying some of the main issues involved. I will frame these issues in terms of several deceptively simple cosmological questions. They are questions that every viable worldview must answer and answer well.

1. *What exactly is the cosmos?* Is the cosmos simply our so-called "physical universe" at any given stage of its existence? Or does it also include other worlds or dimensions, along with any living beings that may inhabit them? In other words, is penultimate reality (i.e., the reality that issues from and depends upon the ultimate reality) comprised solely of what we call the universe, or does it include other kinds of reality as well?

2. *Did the cosmos have a true beginning?* By "true beginning" I mean a moment in time, or at the beginning of time, when the cosmos, in one form or another, *came into being.* In other words, is the cosmos in one form or another eternal, or did it somehow suddenly appear?

3. *If the cosmos did have a true beginning, who or what caused it, and how did it come into being?* Here, we must decide between several possibilities. Did a personal creator draw the cosmos into existence *ex nihilo* (the view of theism)? Or is the cosmos a manifestation, or emanation, of an impersonal mind or spirit (the view of pantheism)? Or is it the case, as some naturalists propose, that the universe sprang into existence altogether apart from divine agency, through a purely naturalistic "quantum fluctuation of nothingness"? Also, if the cosmos did have a true beginning, when did it occur?

4. *When and how did the orderliness that we now observe in the physical universe arise?* Here again we must decide between a few basic alternatives. Is this order eternal—an essential characteristic of an eternal cosmos? Or did it appear more or less instantaneously in a true beginning, presumably at the hand of a divine creator? Or did it arise gradually, as a result of various powers and processes, whether spiritual or physical? In other words, is cosmic order traceable to some kind of evolution or progressive creation?

5. *What is the basic structure of the cosmos?* This question involves several others: Does the cosmos contain multiple worlds or dimensions? What, if any, is the relationship between them? Is there a basic plan or structure underlying the physical universe? Are there moral or spiritual structures in the universe? What is the difference between living and non-living things? What is the difference between man and other living things?

6. *Does the cosmos have a purpose?* This question necessarily involves ascertaining whether the cosmos was created by a

personal god, through whom alone it could have a transcendent purpose. It also requires us to ask how this god might be pleased to reveal to mankind his purposes for the creation.

7. *Can we know with certainty the answers to these cosmological questions?* This much-neglected epistemological question is of vital importance in the debate about origins. Naturalists, pantheists, and theists all speak authoritatively about the beginning, yet their views differ dramatically. Even among leading natural scientists opinions differ, so that one cosmology soon displaces another as king of the mountain. In such a world, we therefore find ourselves asking: can anyone really be sure about the great questions of cosmology? If so, how? What kinds of evidence would be necessary to establish the truth of a particular viewpoint? What, after all, *is* "certainty," and where might it come from? More particularly, is certainty about the beginning possible on any other basis than some kind of revelation from the creator himself?

These, I trust, are the fundamental questions of cosmogony. In the present chapter we will begin to grapple with them by examining and evaluating the answers supplied by modern naturalism. In subsequent chapters we will explore pantheistic and biblical views. I should forewarn you here that all this will take some time. And I believe it deserves time—more, actually, than I can give it. For as we are about to see, in the war of the worldviews, no battles are more fiercely fought than those over the beginning. Seekers must, therefore, be fully prepared to identify all the combatants and to interact intelligently with their arguments and evidences. You know already that in my own hour of testing I was not prepared. I have written what follows in hope that you might be.

Naturalistic Cosmology

In mankind's long history of cosmological reflection, strictly naturalistic views are late, rare, and highly controversial.

They first appeared in Greece, in the sixth century before Christ. Breaking with the mythological cosmology of Hesiod and Homer, the Milesian philosophers (Thales, Anaximander, Anaximenes) and the early Atomists (Leucippus, Democritus) sought to explain the origin and structure of the universe solely by the use of reason and observation.

These men were "monists." They taught that there is a single physical substance underlying the varied objects of our universe. For Thales this substance was water. For Anaximenes, it was air. For Anaximander, it was a mysterious ultimate reality called "the Boundless." For Leucippus and Democritus it was colorless bits of matter called atoms, whirling about in "the void." These early naturalists all agreed that the underlying stuff of the cosmos was eternal and infinite. For them, nature was without a divine creator and without a true beginning.

Nevertheless, said the naturalists, our cosmos (and perhaps others like it) does indeed have a beginning of sorts. This is because the stuff of nature repeatedly congeals, combines, or otherwise shapes itself into various worlds. How, precisely, does this occur? The Milesians, perhaps incorporating spiritual aspects of traditional mythology, believed that nature holds within her bosom an unconscious, impersonal, but dynamic life force by which it forms the ultimate reality into things and worlds. The Atomists, embracing a more severe naturalism, denied the existence of a life force but held that eternally moving atoms come together according to certain necessary laws to produce the world of things. For them, all natural objects and their activities (including the human mind, will, and emotions) are ultimately reducible to the necessary motions of atoms.

Seekers should note carefully that the early naturalists (as well as their Epicurean successors in Greece and Rome) constituted a philosophical minority. They were surrounded by pantheistically minded teachers such as Heraclitus and Parmenides, and were also actively opposed by theistically minded teachers such as Socrates, Plato, and Aristotle. Accordingly, we may safely say that it was not until the 19th and 20th centuries that atheistic naturalistic cosmology actually took a significant hold on the mind of man.

Modern naturalistic cosmology owes much to its ancient counterparts. Nevertheless, in order to understand it in its present form, we must now briefly survey the fascinating and richly significant history that separates the two.[1]

With the birth of Christianity in the ancient world, early Greek naturalism was completely eclipsed by a powerful synthesis of biblical, Platonic and Aristotelian perspectives. The resulting cosmology owed its view of the beginning to the Bible: the triune living God, for His own glory and the good of man, brought the orderly universe into being over the course of six literal days. But this cosmology owed its view of the structure of the universe largely to Plato and Aristotle: planet Earth—with hell deep in its bowels—lies at the center of the cosmos. It is made of four primordial elements: earth, air, fire, and water. Around it, like the layers of an onion, are the seven planetary spheres containing the Moon, Mercury, Venus, the Sun, Mars, Jupiter and Saturn. Beyond these are the three heavenly spheres: the sphere of the fixed stars, the crystalline heaven and, finally, the Empyrean, or "fiery" abode of God. The heavenly bodies themselves, unlike the flawed and changeable earth beneath them, are perfect and eternal, hierarchically arranged, and propelled along their courses by various orders of angels. By God's design these bodies directly influence material objects and historical events on the earth below.

Here was a breathtakingly beautiful picture of the cosmos, a view that encompassed both the spiritual and material realms. To most Christians, this Greco-biblical synthesis seemed like a match made in Heaven. And indeed, the marriage, in one form or another, would last for nearly 1500 years. Who, then, could ever have imagined that epoch-making developments in the 17th century would ultimately lead to an unspeakably painful divorce?

The split was precipitated by the likes of Copernicus, Kepler, and Galileo, all of whom, under the impetus of new discoveries about the heavenly bodies, argued that the sun, rather than the earth, lay at the center of the universe. Such a view spelled big trouble for the reigning cosmology. If it were true, then the earth is in motion beneath the stars, rather than the stars in motion around the earth. And if that is true, then perhaps the stars are stationary and not fastened to

THE MEDIEVAL COSMOS

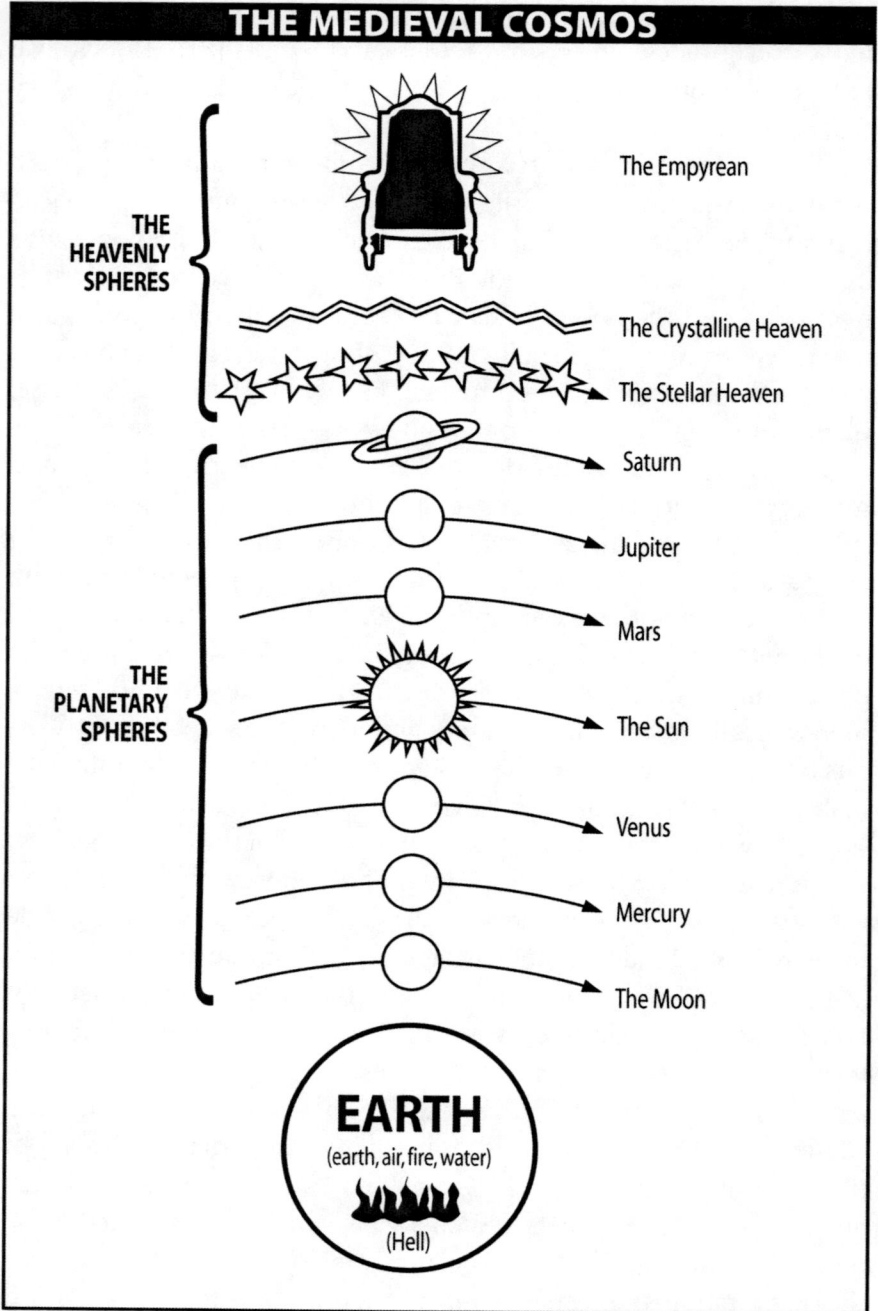

THE HEAVENLY SPHERES

The Empyrean

The Crystalline Heaven

The Stellar Heaven

THE PLANETARY SPHERES

Saturn

Jupiter

Mars

The Sun

Venus

Mercury

The Moon

EARTH
(earth, air, fire, water)

(Hell)

a rotating sphere after all. And if that is true, then perhaps the stars are scattered throughout the cosmic deeps, possibly even throughout an infinitely large universe. And if that is true, where in the cosmos is Heaven, the dwelling-place of God?

Clearly, this new helio-centrism threatened to shatter the entire medieval cosmology. Not surprisingly, therefore, many Church leaders (both Catholic and Protestant) resisted it. It should not be supposed, however, that this resistance stemmed solely from a reluctance to surrender the older view. The more fundamental bone of contention had to do with the proper interpretation of the Bible. Galileo (and other scientists after him) did indeed revere the Bible as a revelation of God's way of salvation. On matters cosmological, however, he felt that God had spoken figuratively, in order to "… accommodate (the truths of cosmology) to the capacities of the common people, who are rude and unlearned." Accordingly, he argued that cosmological truth must be discovered by "… sense experience and necessary demonstrations." In other words, it must be discovered by the methods of natural science alone. In response, Church leaders objected that such an approach impugned the clarity and truthfulness of God's words about creation, and that such an approach to Genesis would eventually undermine our confidence in the rest of divine revelation as well. They therefore opposed Galileo, preferring to trust what they took to be the plain sense of the Word of God over the uncertain and conflicting words of men.[2]

History had reached a crux. A new cosmology was needed. But what role would the Bible play in its formation? Interestingly, not a few of the early modern scientists were determined that its role must remain central. Isaac Newton and Thomas Digges are two good examples. Fixing one eye on the Scriptures and the other on the heavens, they both advanced fresh views of God, Heaven, and an infinite, heliocentric universe. For these men, the demise of classical cosmology was not a threat but a golden opportunity to develop a more truthful—and biblical—understanding of the cosmos.

There were, however, few who followed. Galileo, no doubt unwittingly, had opened a door, and history teaches that under the powerful influence of the Enlightenment many scientists and philosophers now walked all the way through it. Like their Greek predecessors, they discarded the traditional religious view as mere myth. Henceforth, they would take to themselves the twin oars of empirical observation and theoretical reason and launch out into the heavens in search of a new and truly "scientific" cosmology.

The result, not surprisingly, was a burst of cosmological speculation that began in the 18th century and has continued right up to the present day.

Kant and Laplace, in their nebular hypothesis, proposed that god (but not the biblical God) had initially filled an infinite universe with gas that, according to purely natural processes, distilled itself into a host of spinning solar systems like our own. From one of those stars our world was born.

Edgar Allen Poe speculated that there are, in fact, an infinite number of universes, each with its own god who creates them *ex nihilo* as highly condensed particles of matter. The particles explode, expand, coalesce into a cosmos, and finally contract again into nothingness. Here, in exotic form, was an early predecessor of the modern Big Bang hypothesis.

At the turn of the century, Albert Einstein inaugurated the modern era of relativistic cosmology. Like many of his predecessors, he was a true son of the Enlightenment. He did not believe in the transcendent, personal God of the Bible, nor in special revelation, nor in the traditional biblical cosmology. He was, however, a spiritually minded man, describing himself as a pantheist and a believer in "Spinoza's god." Thus, for Einstein, god and the cosmos are one. And since this god is "... an illimitable superior Spirit... a superior reasoning Power (and)... a superior Mind," it is possible for man to grow in his understanding of the universe. True scientific knowledge is, as it were, a gift rising from the depths of the one cosmic mind, to be attained through humble, rational contemplation of the "creation" on the part of man.

Einstein's main contribution to cosmology came early in his career when, in his General Theory of Relativity, he advanced a new understanding of gravity, space, and time. Hitherto, scientists had largely envisioned space as a (infinite) void in which physical bodies were mutually attracted by a force called gravity. While not denying the validity of this model for smaller scales, Einstein proposed that on the largest scales space does not behave like a void, but more like a substance—a substance that is distorted or "curved" by the presence of large quantities of matter, particularly where these are moving at high speeds. On this view, matter influences the curvature of space, while the curvature of space—as a species of gravity—influences the motions of matter.

Einstein and other scientists sought to apply this new idea to cosmology. To do so, they began with two assumptions common in their day. The first was that the universe is *homogeneous*. This means that matter, on the largest scale, is distributed evenly throughout the cosmos. The second is that the universe is *isotropic from all points of view*. Sometimes referred to as *the Cosmological Principle*, this means that the universe looks roughly the same (i.e., spherically symmetrical) to an imaginary observer situated at any vantage point in the cosmos. An important implication of this principle is that the universe has no center, no outer edge, and no place that is unique or "special."

Prior to Einstein, the only way to satisfy the assumed Cosmological Principle (i.e., to avoid a center and edges) was to think of the universe as *open* or infinite. Now, however, with general relativity, the way was open for a new view. For if there were enough matter in the universe, then space may be so highly distorted that it actually "curves back on itself," thereby creating a *closed* universe—a universe that is at once finite, homogenous, and isotropic from all points of view.

It is important to understand that curved space models of the cosmos are, in strictness, unimaginable. To help laymen grasp them, scientists will often propose analogies. For example, they will liken three-dimensional space to the two-dimensional surface of a balloon—a surface that is both curved and stretchable. The scientists admit, however, that these analogies are defective, and that they themselves can only "visualize" curved space mathematically by means of sophisticated four-dimensional geometries. This means that curved space models of the cosmos may be mathematically intuitive (to some), but that the "picture" of the cosmos they yield is counterintuitive (to all), being contrary to our everyday experience and inaccessible to our imagination. More on this momentarily.

Early in his career, Einstein favored the view that our eternal universe is not only closed but essentially "static" (i.e., unchanging in its basic structure), being protected from gravitational collapse by a mysterious repulsive force that he called "Lambda." Later, after the discovery of galactic red shifts, he allowed that the universe may indeed have had a beginning and may be expanding—though not necessarily from an original singularity or as the result of a Big Bang. Thus, to the day of his death, Einstein did not know how, when, or why the universe began.

The view that would dominate twentieth-century cosmology was first proposed in the 1920s by astronomer Georges Lemaitre. He agreed with Einstein that the universe is closed but held that it does not collapse because—as red-shifted light from the stars appears to indicate—it is presently expanding. Moreover, said Lemaitre, this expansion teaches us something about its probable beginning: it points to an explosion of a primeval particle or "singularity" that somehow contained within itself, egg-like, all the stuff (i.e., time, space, and energy-matter) from which the orderly cosmos would evolve over the course of several billions of years. For Lemaitre, the universe itself was eternal and self-existent. Nevertheless, as to its orderly appearance and structure, it had a definite beginning: the Big Bang.

Because the evidence for a Big Bang was slender and equivocal, a lively debate ensued. Parting with Einstein and Lemaitre, Herman Bondi, Thomas Gold, and Fred Hoyle proposed a different model. In 1948 they argued that our eternal universe is "open"—that it is infinite and infinitely expanding. Its eternal uniformity is maintained by the continuous creation of new matter, from which new galaxies are ever being formed. Though no one had ever observed such continuous creation or explained how and why it takes place, the Steady State model of the cosmos enjoyed a significant following for some twenty years.

Then, in 1965, a decisive blow was struck in favor of the Big Bang. George Gamow, a long-time defender of the theory, had for some years predicted that the primal explosion would leave behind a uniform radiation that would reach earth from all directions. In 1965 such a radiation was finally observed, though at a much lower temperature than Gamow had predicted. Since the Steady State model had difficulty interpreting this phenomenon, cosmological opinion swung decisively towards the Big Bang. As we shall see, the Big Bang hypothesis is not without serious problems and formidable opponents. Nevertheless, it has remained king of the mountain ever since.

Characteristics of Modern Cosmology

At the risk of repetition, I will pause here to highlight three important characteristics of modern cosmology. Seekers do well to ponder them with great care.

First, the philosophical tendency of modern cosmology has been away from theism and toward naturalism. More particularly, there has been a three-century drift, first away from the God of the Bible, then from the god of Deism, and finally from any appeal to god at all. This does not mean that all modern cosmologists were atheists (though many were and are). It does mean, however, that ever-increasingly they declined to involve god, miracles, or divine interventions of any kind in their theorizing.

Second, modern cosmology is anti-revelational. That is, its proponents have generally taken the position that mankind cannot receive trustworthy knowledge about the beginning from "religious" sources. This means that we are henceforth shut up to "science." In matters cosmological, scientists and scientific method are the only teachers we are allowed to have.

Seekers should carefully note that this conclusion is not, and cannot be, reached by logic. On the premise that there is no god, it does indeed follow that we cannot receive revelations from him. But how can finite human beings know for sure that that premise is true? And on the premise that there *is* a god, how do we know that he cannot or will not give us revelations – unless, of course, he reveals to us that that is not his way?

But if logic is not the source of this stance, what is? The answer here is complex. Some modern scientists assume the truth of atheism, and therefore automatically rule out divine revelation. Others, like Kant and Einstein, assume special kinds of gods, gods who by definition do not impart knowledge through special revelation, but who do impart it through the right use of reason in philosophy and science. Others, seeing that most religious cosmology is mythological and unscientific, assume that all religious cosmology is mythological and unscientific. Still others, assuming the truth of evolution, conclude that religious cosmologies belong to a primitive stage of man's intellectual development, while "scientific" cosmologies belong to an advanced.

Plainly, there is a lot of assuming go on all around us. Just as plainly, the assumptions have produced a widespread bias against turning to divine revelation for cosmological truth. But are the assumptions valid? Are the conclusions drawn reasonable? Is this bias based on truth? Or could it be based on something less noble? Could it even be based on irrational aversion to admitting the unknown god into the citadel of modern, humanistic science?

Seekers must answer these questions with great care. And as they do, let them fervently hope that the bias against revelation is *not* based on truth. For in our journey thus far we have already seen what many modern scientists have not: natural science has neither the calling nor the competence to take us to the beginning. Nor can it answer any of the other questions of life. Nor can philosophy. If, then, divine revelation does not come to our aid, we are in the dark forever.

Finally, modern cosmology is largely committed to evolutionism. It tends not to view the universe as a static creation (as did the medieval mind) but as a continually evolving or devolving reality. It is, as it were, a growing, living, and dying material organism. This perspective, which was first seen in ancient Greece, fully emerged in the 19th century when evolutionary geology (spearheaded by Charles Lyell) and evolutionary biology (spearheaded by Charles Darwin and Julian Huxley) won broad allegiance among the western intelligentsia. In time it overflowed into other areas of inquiry. Eventually, the idea of *cosmic evolution* so firmly took hold upon the Western imagination that today many treat it as an axiom and the organizing principle of all science and culture.

A few citations will demonstrate this important point.

Evolutionist Julian Huxley, tracing the history of evolutionism, wrote as follows:

> The concept of evolution was soon extended into other than biological fields. Inorganic subjects such as the life-history of stars and the formation of the chemical elements on the one hand, and on the other hand subjects like linguistics, social anthropology, and comparative law and religion, began to be studied from an evolutionary angle, until today we are enabled to see evolution as a universal and all-pervading process.[3]

The pantheistically minded Teilhard de Chardin agrees:

> Evolution is a general postulate to which all theories, all hypotheses, all systems must henceforward bow, and which they must satisfy in order to be thinkable and true. Evolution is a light which illuminates all facts, a trajectory which all lines of thought must follow.[4]

These are strong, even imperious words, echoed today by an intelligentsia that often ridicules anyone who dares to question the evolutionary paradigm. Careful seekers, however, will not bow to evolutionism simply because men assert that it is true or picture it as such. They will bow only if the evidence proves it to be true.

Summary

All of this brings us at last to a brief statement of the modern, naturalistic cosmology. In setting it forth, I will incorporate the naturalist's responses to many of the basic cosmological questions cited at the beginning of this chapter.

According to most modern naturalists, the cosmos, in one form or another, is eternal. About 15 billion years ago it existed as a super-condensed "singularity" containing all the time, space, and mass/energy of the universe. For reasons not as yet understood, it exploded. Then, after a very rapid initial expansion, light elements such as hydrogen and helium were formed from energy and subatomic matter. By gravitational attraction, these coalesced into primitive stars and galaxies. Inside the stars, nuclear reactions generated carbon and oxygen. In time a second generation of stars was born from the first, in which heavier elements were formed. In this process, there somehow arose in our neck of the woods a solar system of several planets orbiting a star we call the sun. In still more time, one of those planets somehow evolved water, a primitive atmosphere, organic molecules, and plant and animal life. Finally, the whole process produced man, whose remarkable brain chemistry exuded the mystery of consciousness. And thus, quite apart from any intelligent creator, the universe was suddenly able to look upon itself and think about its origins. In man, the cosmos was now able to contemplate its own beginning.

Interestingly, the Big Bang cosmology is currently embraced not only by naturalists but by many pantheists and theists as well. This is hardly surprising since the existence of the primeval particle, its sudden explosion, and the wondrous development of raw energy into a highly complex and ordered material cosmos all cry loudly for the participation of an intelligent and powerful supreme being.

Nevertheless, the strict naturalist does not (or will not) hear the cry. For him there is no god, no spirit (whether animal, human, or angelic), no spiritual realm, and no life force—in short, nothing supernatural. Nature, in one form or another, is all there is, eternal and uncreated. Accordingly, behind its evolutionary development there is no divine intelligence, purpose, plan, or activity. Rather, the material universe—by a mysterious combination of explosion, mere chance, rigorous natural law, and (in the case of life) random mutation and natural selection—somehow shaped itself into the orderly cosmos that we observe and contemplate today. It is a miracle without a miracle worker; the product—in Richard Dawkins' famous phrase—of a "blind watchmaker."

Naturalists freely admit that this view strains credulity, that it has worthy opponents, and that it is not without serious theoretical and observational problems. They even concede that we, who were not present in the beginning, can never really know for certain whether it is so. But since they know there is no god, they also know that something like this *must* have taken place. And until they can come up with a better explanation they are pretty much agreed in presenting this one to the world as the truth.

Evaluating Naturalistic Cosmic Evolution (NCE)

Many who have traveled with me thus far are now in pain. Or at least they should be. In the earlier summary of *The Test* they were introduced to an impressive body of evidence suggesting that the Bible is a trustworthy revelation from the unknown god and that Jesus is his appointed Teacher. This means, of course, that both the panoply of signs and the testimony of Jesus lend supernatural authority to the core affirmation of biblical cosmology: Yahweh Elohim, the LORD God of Israel, created a mature, fully-functioning universe in six days, pronounced it good, and then "rested" from His creative work. Yet throughout their lives they have also repeatedly encountered what appears to be a large body of scientific evidence supporting cosmic evolution and have themselves become part of a widespread cultural consensus to the effect that cosmic evolution is an established scientific fact. In short, the two cosmologies seem well supported by good evidence; the two are completely irreconcilable; and the two, crashing together in one head, therefore, produce pain.[5]

I have felt it. Back in 1970, as I studied with Father Barry and seriously considered the Bible for the first time, I knew immediately that both cosmologies could not be true and that at least one of them was a myth. How did I resolve the tension? I am afraid that with an educational experience such as mine, and with no one in my life to defend the biblical view, the result was all but inevitable.

Today, however, the situation is much changed. Yes, seekers are still in conflict over competing views of the beginning. But now the pendulum appears to be swinging in the other direction so that many seekers are actually quite open to seeing cosmic evolution as the myth and the Bible as the truth. Evolutionism remains, of course, the dominant cosmological model among the Western intelligentsia. And it is still purveyed as fact in our schools and media. But even the least alert among us realize that the theory of evolution has become highly controversial, that it is now "a theory in crisis."[6] As we are about to see, this is because the traditional evidences and arguments for cosmic evolution have been largely discredited, and because much new evidence favors not only divine creation, but divine creation of the biblical kind.

As a result of all this, some observers are even suggesting that the signs of a cosmological "paradigm shift" are appearing all around us. Many scientists—both religious and nonreligious—have begun privately to express doubts about cosmic evolution. A few, daring to break with orthodoxy, have openly exposed its problems to the lay public.[7] The Intelligent Design Movement is now gaining a respectful hearing in our universities.[8] The older Biblical Creationist Movement is now much enlarged, strengthened, and streamlined by a new generation of zealous young scholars. Quality creationist books, magazines, and films are available to the public. Sophisticated creationist websites attract millions of visitors yearly.[9] Parents throughout the country are urging school districts to let their children hear both sides of the great debate.

Could it be, then, that the thought-world of my own childhood is about to be turned upside down? Could it be that tomorrow's children will look upon cosmic evolution as "the great cosmogenic myth of the twentieth century"?[10] Could it be that generations to come will view one or another form of divine creation as well established scientific fact?

Time will tell. In any case, this much is sure: a great debate about the beginning is currently in progress. To thoughtful seekers it speaks of a test. Desiring to pass it, they also desire carefully to examine the evidence on both sides. Only then will they decide what is myth and what is truth.

We must turn, then, to an evaluation of naturalistic cosmic evolution. In the space allotted to me I cannot, of course, do full justice to a controversy that now engages the attention of thousands of scientific professionals, many of whom have written extensively on this subject. Nevertheless, because of its great importance it is vital at least to touch on the main issues and arguments involved. Beyond that, I must rely on copious endnotes to direct motivated seekers to valuable books, videos, and websites where they may plumb the depths of the great debate to their heart's content.

Is NCE Intuitive?

Is the naturalistic cosmogony intuitive? To judge from the ninety percent of Americans who reject it, apparently not.[11] This rejection stems not only from its inherent atheism but from the many cosmological propositions that necessarily flow from it. Let us consider a few.

First, NCE asserts that the cosmos, in one form or another, is eternal. It would have to be since, on naturalistic premises, there is no god to create it. But for whatever reasons, most of us balk at this assertion. We are comfortable enough ascribing eternity to a self-existent creator spirit but ill at ease ascribing it to the material cosmos itself. Intuitively, we sense that the physical cosmos is a *dependent* reality—that its very existence, as well as its marvelous order and proper functioning, all depend upon a distinctly spiritual reality higher than themselves.

NCE also denies the existence of any lesser supernatural realities such as spiritual realms, angels, demons, animal or human souls, or a nonphysical life force animating matter. On atheistic premises, this denial makes perfect sense: where could such realities come from, if not from a spiritual creator? But again, most people do not accept the premises. Therefore, the naturalist's denial of lesser spiritual realities—and especially of the human soul—strikes them as counterintuitive.

Modern NCE declares that "in the beginning" all the time, space, and mass/energy of the universe was bundled up in a primeval particle. Such a proposition, to the extent that we can even conceive it, is highly counterintuitive. Yes, the *measurement* of the motion of a body is inseparable from considerations of time and space (e.g., the miles-per-hour it travels). But can we really leap from this simple fact to the conclusion that time and space themselves are inseparable from matter, or that they were once compressed together, substance-like, in a singularity?

But let us consider these matters somewhat more closely, beginning with the naturalist's idea of time.

Time, we are told, is an altogether natural phenomenon that began with the Big Bang and is strictly tied to motion—whether of light, objects in space, or (the stretching of) space itself. But common sense clearly contradicts this view. Observe, for example, the workings of your own mind as you try to understand the Big Bang: though warned against it, you simply cannot keep yourself from imagining the singularity suspended in time and space, or from pondering how long it was there before it exploded, or from asking how it got there in the first place, etc. Indeed, in their unguarded moments, naturalists themselves address these very questions! Why? Because we all have the same intuitive understanding of time: it is eternal, going infinitely far back into the past and infinitely far forward into the future. Therefore, try as we may, we cannot accustom ourselves to the idea of time having a beginning.

The more we ponder time the more we also realize that it has a subjective, or spiritual, dimension. In other words, time essentially involves consciousness. In particular, it involves a *consciousness of duration*—an ineffable awareness in some person's mind that he and the world he lives in are moving out of the present, into the future, leaving behind the past.

To get a feel for this point, you may perform a humble thought experiment. Picture a rock in your back yard. You know intuitively that the rock is not conscious. Therefore, for the rock, time does not exist. This does not mean, however, that the rock does not exist in time. For if you are conscious of the rock, then it exists in time, since you are *aware* of its existing in time. What's more, you know that even if you are not conscious of the rock, it still exists in time. And reflecting on this, you finally realize that it continually exists in time only because there is a continuing god who is continually conscious of it.

Time, then, as opposed to rates of physical change, is *essentially* related, not to the speed of light or to the motion of physical bodies, but to consciousness and self-awareness. In other words, it is not a natural but a *supernatural* phenomenon. Indeed, it may be that we humans find ourselves "in" time, not because we are in the universe but because we (along with the universe) are in god and because he allows us to share a limited awareness of his own eternally enduring self. If this is so, it is hardly surprising that statements about time "beginning" with the Big Bang—with neither god nor man upon the scene—scandalize our intuitions.[12]

The modern view of space is, if possible, even more unsettling. Big Bang cosmologists speak of space as though it were a physical substance or an attribute of matter: at one time all space was "compressed" into the primeval particle; after the Big Bang it "stretched" or "expanded" (along with time and matter) to produce our present universe. Furthermore, if we are thinking of this space as expanding away from a center, we are getting it wrong. For, according to the Cosmological Principle, space has no center, no outer edge, and no fixed frame of reference. Rather, it is something like the surface of a balloon with dots on it: it somehow rests upon yet another reality (or dimension) which, having exploded, stretches it, thereby causing the objects forming and moving in it (i.e., "dots" of matter, galaxies, stars, etc.) to move away from one another.

Again, it is important to understand that these torturous ideas were not generated by astronomical observations. Indeed, as we shall see later, astronomical observations frequently militate against them. Rather, they were generated, first by exotic, non-Euclidean geometries, and secondly by a philosophical bias against traditional Greek and biblical views which placed the earth at or near the center of the universe. Observe how this bias is reflected in the words of cosmologist Edwin Hubble, who wrote, "The unwelcome supposition of a favored location (for our earth or galaxy) must be avoided at all costs (and) is intolerable."[13]

But tolerable or not, most people find the ancient view far more comprehensible—and therefore more attractive—than the modern. Intuitively we reckon space to be a static, immaterial reality that could easily exist independently of matter and that could readily be filled by matter to a greater or lesser degree. Space, for most us, is rather like the unknown god—an unchanging element or medium *in which* changing material objects live and move and have their being. Furthermore, if it were one day shown that the universe is, after all, a vast sphere whose center is at or near the earth, the common people would doubtless hear it gladly. The same cannot be said, however, for the strange and imponderable universe of the modern naturalist.

But perhaps most counterintuitive of all is the naturalistic conception of the history of matter. NCE asserts that our orderly and (in some parts) fantastically complex cosmos is actually the end result of an explosion of raw energy, subsequently wrought upon by nothing more than (lots of) time, "chance," and the impersonal laws of physics.

This scenario offends intuition at many points. Since when did an explosion ever produce anything but disorder? Why would raw energy, contrary to its observed behavior, congeal into matter rather than simply dissipate into space?[14] Why would it distill itself, not into one but into hundreds of different molecules and compounds? And how could these in turn distill themselves into innumerable discreet objects, intricately and (in some cases) vitally related to others? Yes, such a journey from chaos to cosmos could be accomplished by the hand of a god working upon exploding matter (hence the appeal of theistic evolution). But to suggest that it is accomplished by exploding matter itself is so counterintuitive as to invite charges that someone is purposely resisting the obvious in order to avoid reckoning with an infinitely powerful and intelligent creator.

And what of the naturalistic idea of life? The naturalist assumes that physical matter/energy is all that exists. He therefore concludes that living beings must be aggregates of inorganic matter that have somehow crossed a certain threshold of complexity so as to be animated by a certain form of physical energy (i.e., electro-chemical).[15] But such a view frustrates common sense. Yes, most of us are prepared to admit that organic life *involves* matter and energy in special arrangements and kinds of motion. But few of us will admit that that is *all* it involves. When, for example, we observe our pet cat stalking a bird, or arching its back beneath our outstretched hand, or circling our feet at dinnertime, we do not naturally ascribe the motions of her body to firing synapses and twitching muscles. No, we ascribe them to some kind of metaphysical principle animating the cat, whether her "soul" or something beyond her soul. This is the intuitive view of life—that it "rides" on chemistry and takes hold of chemistry but that it cannot be reduced to chemistry alone.

Naturalists, of course, deny the existence of a supernatural life force. But for most of us, life is inconceivable apart from such a force. When this force is present, physical bodies grow, move, eat, reproduce, play, work, etc. When it departs—no matter how complex they may be—the bodies die. We may not know how to explain all this, but we do know that life is something more than matter. It is a supernatural something, worthy of deep reflection and profound respect.

Finally, we have the naturalistic view of man. For the naturalist, a human being is essentially a highly evolved organism, and the human mind— so serviceable in the organism's struggle for survival—a byproduct of the electrochemical activity of the brain. But again, while most of us will admit that mind and brain somehow work in tandem, very few will agree that mind can be "defined away" as brain activity alone. Why? Because we understand intuitively that the two realities are heterogeneous. The brain is a physical something, while the mind is a spiritual or metaphysical something. We can observe and handle the brain, but not the mind. Unlike the brain, the mind cannot even be located in time and space. The mind is not a part of nature at all. In fact, it is so supernatural that it can somehow take all of nature into itself, most especially when it is doing cosmology![16] Knowing all these things to be true, we therefore quite naturally tend to resist naturalism's understanding of man.

We find, then, that in many ways NCE proposes a highly counterintuitive version of the cosmos and its beginning.

Is NCE Reasonable?

Is NCE reasonable? In other words, is this hypothesis logical and well supported by good evidence? The discussion that follows, which cites key scientific issues in the modern debate about origins, will show that it is not—and why theories of cosmic evolution are now so controversial. The material is categorized under three main headings, with numerous illustrations cited under each. Note carefully that much of the evidence speaks not only against naturalistic evolution but against theistic and pantheistic evolution as well.

NCE Violates Natural Law

Since the advent of modern science some 400 years ago, our understanding of the laws of nature has dramatically increased. Here are just a few of the well-established natural laws that conflict sharply with NCE.

1. **The Law of Cause and Effect**

   According to this law, every event or state of affairs is an effect that has a cause adequate to produce it. NCE teaches that a primeval explosion, (lots of) time, chance, gravity, spontaneous generation, random mutation, and natural selection all worked together on matter to produce the stupendous order and complexity of our cosmos. However, modern science itself has shown that none of these causes is adequate to do so (see below), while it is self-evident that a divine creator is. This implies that order in the universe must be traceable to the active involvement of a powerful, supernatural intelligence. More particularly, the law of cause and effect requires that the mysteries of life, consciousness, and personality have a living, conscious, and personal cause. Dead matter is simply not adequate for the job. A living, conscious, personal god is.

2. **The Law of Universal Gravitation**

   According to this law, material objects are attracted to one another in inverse proportion to the square of their distance. This means that the closer they get, the more powerfully they are drawn and held together. If, therefore, all the mass of the universe were concentrated in a tiny singularity, the gravity holding it together would be stupendous. It would be a super-massive black hole. No natural force known to man could overcome such gravitational attraction. How, then, could the singularity explode?

3. The Second Law of Thermodynamics

According to this law, there is a natural tendency in all systems (e.g., stars, planets, rocks, cells, plants, people) to progress from order to disorder through the loss of energy available for their preservation and/or transformation. In other words, as time marches on things lose heat and run down. The Second Law, therefore, encourages us to think of our orderly cosmos as having a definite beginning and as gradually moving from a highly ordered state towards a disordered state—exactly what we observe in nature. NCE, on the other hand, asks us to think of the cosmos as beginning in a highly disordered state (i.e., chaos) and moving towards a more ordered state—exactly what we do not observe in nature.[17] The conflict is stark and recognized by all, friend and foe alike.[18]

There is, of course, an apparent exception to this principle—the case of living beings—but it actually scores important points against naturalism. Yes, for a season living beings do grow, develop, and increase in complexity, thus (temporarily) defying the Second Law. But this occurs only because highly complex, pre-existing energy conversion mechanisms (e.g., reproductive cells loaded with genetic instructions) are *already* in place, transforming food into chemical energy and directing the development and proper functioning of various molecules, organs, and living systems. Furthermore, as we have already seen, it is not really the cells or organs that produce life but life that produces (and animates) the cells and organs. The case of living beings, therefore, teaches us that cosmic evolution is inconceivable apart from the activity of a supernatural life force working through some pre-existing cosmic mechanism (analogous to cells and genes) by which raw energy from the Big Bang might have been transformed and built up into a complex functioning system. But NCE denies the former and cannot find the latter. Its version of cosmic evolution is, therefore, impossible.

It is also well worth noting that among living things the Second Law operates not only at the point of death but also upon the mass of genetic material that shapes life. For example, we know that over the years the total gene pool of earth's living beings has significantly decreased through the extinction of many life-forms. Also, we have learned that in the transmission of genetic information from one generation to the next, there are ever-accumulating "failures of communication" (e.g., mutations, copying errors, etc.) that weaken or injure the offspring. This means that, given enough time, the Second Law would render all life extinct through the progressive decay of its genetic base.

These important biological facts speak powerfully against NCE. In particular, they imply that life is actually *devolving*, rather than *evolving*; that in the beginning it was healthy, but now is being pushed by the Second Law toward sickness and death; that in the beginning there must have been a process of creation that now, for some reason, has ceased and given way to a process of destruction. Such a scenario is very far from the modern naturalistic worldview—and very close to the biblical.

It appears, then, that the Second Law of Thermodynamics, with its many profound cosmological implications, threatens to drive a stake into the very heart of NCE.

4. The Law of Biogenesis

More than a hundred years ago Louis Pasteur first articulated this fundamental law of biology—that life always comes from life; that spontaneous generation, or life arising from non-life, never occurs. NCE teaches, however, that 1.5 billion years ago a single living cell spontaneously arose in a nonliving "prebiotic soup." No scientist has ever discovered such a soup, nor observed the spontaneous generation of a single cell, nor produced one under ideal laboratory conditions. Yet many assert dogmatically—though quite unreasonably—that "in the beginning" spontaneous generation really did occur.

5. The Laws of Probability

Careful observation shows that the so-called "chance" events of nature are actually governed by laws of probability. As casino owners well understand, knowledge of these laws enables us to make predictions about what events may reasonably be expected to occur. What, then, is the probability that a single protein, gene, or cell might "spontaneously" form out of the random molecules of a pre-biotic soup? Because we now know that these "simple" building blocks of organic matter are so fantastically complex, mathematicians have repeatedly concluded that the probability is so small as to render such events impossible.

Sir Fred Hoyle, for example, calculated that the likelihood of "life" forming from inanimate matter is $1 \times 10^{40,000}$. Dr. Emile Borel, who discovered the laws of probability, wrote that an event whose chances of occurring are beyond 1×10^{50} is an event that will not occur, no matter how much time or how many trials are allowed. Dr. Edwin Conklin, himself an evolutionist, concludes, "The probability of life originating by accident is comparable to the probability of the unabridged dictionary resulting from an explosion in a printing shop." And all this is to say nothing about the probability of millions of different life forms accidentally evolving from the one original cell. Is it then reasonable for NCE to insist that such events actually took place?[19]

It is worth noting that this powerful argument does have a flaw, but a flaw that actually strengthens the case against NCE. This is because it grants to the naturalist his premise that the mere juxtaposition of simple molecules might be sufficient to "create" a (complex) building block of life. But in a universe governed by the Second Law of Thermodynamics, this is not possible. Simple molecules cannot simply *come* together to produce proteins, genes, or cells. Rather, they must be *brought* together and *wrought* upon. In other words, a powerful intelligence must overcome the Second Law to create a complex organic base for life. Furthermore, we must again remember that life involves more than organic complexity. A living cell, as I have argued, is more than a functional assemblage of organic molecules. It is an assemblage assembled by life, held together by life, and animated by life. It is a metaphysical as well as a physical entity. The naturalist, however, denies the supernatural, by which alone organic evolution could occur. Accordingly, even if there were an infinite number of naturalistic universes filled with simple molecules (as some naturalists now propose), none of them would ever produce a living organism. The probability of this event is not very, very small. It is zero.

The complex orderliness of living things, so recently unveiled by biological science, has roused many a modern skeptic from his naturalistic slumbers. An outstanding case in point is Dr. Anthony Flew, arguably the most influential atheistic philosopher of the 20th century. Having rejected all the classical proofs for the existence of a god for over 60 years, he finally found, at age 81, that the complexity of life (and especially of the DNA molecule) compelled him to embrace the theistic position. Labeling the naturalistic scenario for the origin of life "improbable," Flew concluded,

> Science has shown, by the almost unbelievable complexity of the arrangements that are needed to produce life, that intelligence must have been involved. I have been persuaded that it is simply out of the question that the first living matter evolved out of dead matter and then developed into an extraordinarily complicated creature. My whole life has been guided by the principle of Plato's Socrates: follow the evidence wherever it leads. The conclusion is—there must have been some intelligence.[20]

6. The Law of Genetic Stasis

Gregor Mendel, the father of genetics, discovered that generational variation in living beings normally occurs not because different genes are formed in offspring but because identical genes in parents combine in different ways. For example, the parent genes of a sweet pea may combine to produce an offspring that is red, pink, or white; larger or smaller; more or less fragrant, etc. But there are well-defined limits to such changes. The sweet pea's genes cannot combine in such a way as to produce a rose or a rabbit. In other words, Mendel discovered that the various "kinds" of life are essentially static. Their gene pool allows for small variations (sometimes called microevolution), but not for big changes that would alter their identity (sometimes called macroevolution). This truth is borne out by animal breeding experiments, and also by the fossil record of past life forms. NCE, however, asserts that macroevolution does indeed occur—that one "kind" gradually changes into another because of numerous random mutations in the genes. The theory, therefore, violates Mendel's well-established law of genetic stasis.

7. The Law of Irreducible Complexity

Ever-increasingly, biologists have come to realize that organic life is composed of complex systems. One important characteristic of these systems is that they are "irreducibly complex." That is, they require a minimum number of components in order to function properly. Cells cannot survive without their DNA, membranes, mitochondria, nucleus, etc. The human eye cannot function without the lens, retina, optic nerves, or brain centers for sight. The complex navigational systems of dolphins, bats, birds, and insects would be useless if all their components were not present and operative. The philosophical implications of irreducible complexity are weighty and only now coming into full view. Irreducible complexity means that biological systems *must* have been intelligently designed, and that they *must* have sprung into being fully formed and functioning. How could the systems function and the organisms survive if they had not?

Naturalists, of course, reject this (creationist) conclusion. But their view—that such systems evolved piecemeal by way of small mutations—is impossible. This is because small mutations would correspond to the appearance of "incipient structures" in an already viable organism—a tenth of an eye, a half of a wing, a proto-leg, etc. The problem here is that such structures would definitely be useless and probably be deadly. In the competition for survival, creatures with useless and burdensome incipient structures would be at a disadvantage. They would quickly be "selected out" and perish, while the population of "normal" parents and offspring (whose existence is inexplicable on naturalistic grounds) would endure. It appears, then, that NCE cannot be reconciled with the phenomenon of irreducible complexity, while sudden, theistic creation actually predicts it. Which of the two, then, is more reasonable for us to believe?[21]

NCE Is Not Supported by Good Evidence

Down through the years evolutionists have cited a number of evidences in favor of their theory. In the following survey, we look at a few of the more important ones. As we do, we will discover one of the main reasons why NCE is now so controversial: the evidences advanced to support it are either equivocal or now falsified.

1. **Galactic Red Shifts**

   Astronomers have observed that light from the galaxies is usually (but not always) shifted towards the red end of the spectrum, with what are presumed to be the most distant galaxies showing the greatest amount of red shift. Here Big Bang cosmologists find support for their theory. They argue that the light is "stretched" primarily because space itself is expanding, and that space is expanding because of a Big Bang.

There are, however, some problems. For example, certain objects with high red shifts (e.g., galaxies or quasars) are observationally connected with other objects of low red shifts (e.g., other nearby galaxies). But if these objects are really traveling at such different speeds, and if red shifts correlate with distance from an observer, how can the objects be close to each other?

Also, if red shifts are the result of objects accelerating away from us, we would expect measurements of the shifting to be "smooth"—spread out evenly over a wide range of numerical values. In fact, however, it is now known that the red shifts are "quantized"—that they cluster around specific, regularly spaced values. This has led some astronomers to propose other explanations for red shifts. One is that the stars and galaxies are arranged in expanding shells—a view that entails the shells having the earth or the Milky Way at or near their center. Another, and the most popular, is the "tired light" view—that red shifts occur as photons interact with one another and lose packets of energy during their journey through space. If this view is correct, it means that the universe may not be expanding at all.

In sum, red shifts do not prove that the universe really is expanding. And even if it is, this does not necessarily prove that a Big Bang has caused it to expand. A Big Bang is the favored interpretation for galactic red shifts, but it is neither the only interpretation nor the best.[22]

2. Cosmic Background Radiation (CBR)

Astronomers have observed a uniform radiation of heat (2.73 degrees K) coming to us from all directions. Big Bang cosmologists argue that this radiation is a "snapshot" of the universe shortly after the Big Bang when energy/matter had not yet clustered into galaxies and stars but was uniformly distributed in space. The problem here, however, is to explain how uniformly distributed energy/matter could so quickly congeal into heavenly bodies, bodies which are not at all evenly distributed in our exceedingly "clumpy" universe. Because these difficulties are so acute, some astronomers propose other explanations for the CBR: that it is the "temperature of space" produced by starlight; that it is the energy residue of tired light; that it comes from the magnetic field or cosmic rays of our own galaxy, etc. Though differing among themselves, these scientists agree that the CBR is far better explained by something other than a Big Bang.[23]

3. Fossilized Geological Column

As we learn from places like the Grand Canyon, the surface of the earth in many places is layered. These layers of sedimentary rock often contain the fossil remains of dead creatures. As a general rule, we find simpler creatures in the lower layers and more complex creatures in the higher. Darwin and his followers argued that these strata were laid down gradually over millions of years and that the fossils prove that life evolved slowly from simple to complex forms. Today, however, more and more scientists are willing to admit that the so-called geological column not only fails to demonstrate evolution but that it has become a positive embarrassment to the theory. The reasons for this are many.

First, there are clear indications that the strata were laid down suddenly and catastrophically. These include mass graves, fossils of fish eating other fish or giving birth, fossils that pass vertically through several layers (e.g., trees), contiguous layers of different material with no evidence of erosion between them, upwarped parallel strata indicating that the layers were soft and pliable when bent, etc.

Second, out-of-place fossils repeatedly contradict the standard evolutionary scenario. For example, land animals, flying animals, and marine animals have been found side by side in the same rock; horse footprints have been found in rocks dating to the era of the dinosaurs, etc. On evolutionary premises, all these life forms should be separated by millions of years.

Third, the evolutionary tree has no trunk. In other words, there is no evidence that multitudes of complex life forms evolved from a few simple life-forms. Instead, fossilized life appears "… suddenly, full-blown, complex, diversified, and dispersed world-wide."[24] The lowest life-bearing layers (i.e., the Cambrian) include fish, worms, corals, plants, and even some vertebrates. Beneath them there is nothing at all. Tellingly, the Cambrian layers also reveal *more* life forms than exist today, while the standard evolutionary scenario predicts fewer.

Finally, there are the notorious "gaps" in the fossil record. If evolution were true, we should be able to find millions of transitional forms bridging the gap between the basic kinds of life (e.g., forms linking bacteria to plants, plants to fish, fish to amphibians, amphibians to reptiles, reptiles to birds and mammals, cows to whales, apes to men, etc.). After 150 years of studying billions of fossils, scientists now know that such transitional forms simply do not exist. The late Stephen J. Gould wrote, "The extreme rarity of transitional forms in the fossil record persists as the trade secret of paleontology. The evolutionary trees that adorn our textbooks have data only at the tips and nodes of their branches: the rest is inference, however reasonable, and not the evidence of the fossils." Here we have what is arguably the single most important piece of empirical evidence against the theory of biological evolution. Why do the textbooks not speak of it?

In sum, the geological column supplies no solid evidence for NCE, and much against it. It does, however, harmonize well with the biblical claim that all kinds of life were simultaneously created "in the beginning" and later destroyed in a global flood that both quickly laid down the geological column and completely transformed the face of the earth.[25]

4. Homologous Structures

Living beings have similarities. All of them begin as simple cells whose growth is determined by their DNA. Many are generated sexually and pass through similar stages, from zygote to embryo and onward. Many have structural similarities: skeletons, nervous systems, respiratory systems, circulatory systems, etc. Evolutionists argue that this "structural resemblance signifies blood relationship." In other words, evolutionists contend that common (or homologous) structures are evidence of common origin.

It is clear, however, that common structures could just as well point to a common designer. Indeed, this is the better alternative since we may reasonably suppose that a single creator would want to teach us that he is the sole author of all forms of life by using a single basic plan (e.g., genetic coding) modified by creative variations. Evolutionists, on the other hand, posit no such designer and therefore cannot explain why living things should have the similarities they do. Their theory of random changes predicts a "chaos" of life-forms, whereas what we actually observe is a "cosmos" of life-forms—an ordered family of living beings characterized by unity and diversity of structure.

Also, it is worth noting that homologous structures often speak against common origin. This is especially evident at the molecular level. For example, crocodile hemoglobin is more similar to chicken hemoglobin than to that of snakes and other reptiles. The cytochrome-c structure of a rattlesnake is closer to that of a man than to that of a snapping turtle or bullfrog. Human lysozyme (an enzyme for digesting food) is closer to chicken lysozyme than to that of any other mammal. If similarity of (chemical) structures indicates common origin, then on the traditional evolutionary "tree" crocodiles should be closer to chickens than to snakes, rattlesnakes should be closer to men than to turtles, etc. But they are not. Perhaps, then, the evolutionary tree is in error. And perhaps observed

similarities in chemistry are actually based on a common designer who has incidentally supplied the world with powerful
evidence against the hypothesis of common ancestry![26]

5. Microevolution

Scientists have observed minor adaptive changes in the
structure of various organisms. Darwin's finches, peppered
moths, pesticide resistant mosquitoes, as well as the varieties
of plants and animals bred by man—all are cited by naturalists
as proof that living things evolve. But such claims involve a
serious misunderstanding. Evolution, by definition, means *the
appearance of more complex organisms from simpler organisms
due to an increase in genetic information.* In the cases just cited
there is no such increase, only a reshuffling of information,
or possibly a loss of information due to mutation. In other
words, so-called "microevolution" is not evolution at all but
simply *variation* within a pre-existing kind of life. Variation
is observed—and even intentionally produced—all the time.
Evolution is never observed, and has never been produced.
Therefore, "microevolution" is not solid evidence for NCE. It
can, however, readily be seen to testify to the goodness and
ingenuity of a divine creator who provides for the survival
of specific "kinds" by endowing them with a *limited* genetic
capacity for adaptation.

6. Hominid Fossils

Nothing has more powerfully promoted faith in evolution
than artistic reconstructions of "early man." It is, then, all the
more shocking when serious inquirers discover that there is
little or no good fossil evidence to support them. The fossils of
so-called "hominids" (man-like precursors to *homo sapiens*) are
few and fragmentary (all of them would fit nicely on a single
pool table). In some cases they have been exposed as frauds
(e.g., Piltdown Man). In others, they have been identified as
(extinct) apes (e.g., *Ramapithecus*, various australopithecines,
and the suspicious Java and Peking Man). In still others, they

are now seen as fully human (e.g., *Homo Erectus* and Neander-tal Man). It is true that some paleontologists still argue that Lucy and her kin (*australopithecus afarensis*) were transitional. Others, however, (including her discoverer, Donald Johanson) assert that she is too different structurally to be related to man, though she is very much like a pygmy chimpanzee or bonobo.[27] It appears, then, that hominid fossils actually confirm what we see around us and have learned from the study of genetics: despite notable similarities, there is a vast chasm between apes and men, with no known intermediate forms to bridge the gap.[28]

7. Outmoded Arguments

Because of scientific advances over the last 150 years, many evidences and arguments for evolution have been discarded. Unfortunately, some of these still appear in high school and college textbooks. We must, therefore, mention them briefly.

Vestigial Organs: Older evolutionists cited over 150 apparently useless structures as proof that evolution is presently occurring, bringing in new, more functional structures even as it casts off the old. Today we know the function of nearly every one of them, (e.g., appendix, thyroid gland, pituitary gland, thymus gland, pineal body, coccyx, tonsils, wisdom teeth, etc.). Furthermore, even if these structures were gradually disappearing, they would only prove *devolution* (i.e., a loss of genetic information and its corresponding structures), not evolution (i.e., an increase of the same).

Laboratory Life: In the famous Miller-Ulrey experiment, an electrical charge passing through a mixture of gases produced simple amino acids. Since amino acids are the building blocks of proteins, and proteins are the building blocks of "life," it was felt by many that science was not far from synthesizing life itself. Today we know better. Scientists cannot even synthesize complex amino acids, let alone proteins or the fabulously complex universe of a single living cell. And even if they could, it would only prove that intelligence is required to produce them. Also, we must not forget that life is more than the physical structure that carries it. It is only when "life" lays hold of a physical carrier that we have a living being. Therefore, even if scientists could create the carrier, they would still have to create "life" and infuse it into the carrier. Life, it would appear, requires the active presence of a living god. Mere human beings—even the smartest among us—cannot fill the bill.

Embryology: Years ago, evolutionists held that developing embryos traverse the entire course of their evolutionary history, that "ontogeny recapitulates phylogeny." A common proof for this assertion was the so-called "gill slits" of mammal embryos, purportedly showing their ancestry from fish. Today we know that the "slits" are actually folds that develop into various parts of the mammal's face and head. Scientists have discovered no useless, purely "commemorative" stages in the development of living beings. Furthermore, if some embryos temporarily have similar appearances, why do such similarities (which are not reflected in their genetic makeup or early stages of development) prove a common evolutionary ancestry? Surely a more reasonable conclusion would be that they were created by a common designer.

The Races of Man: Early evolutionists, seeking to contradict the biblical teaching that all mankind descends from Adam and Eve, argued that the existence of different "races" (e.g., Caucasoid, Mongoloid, Negroid and Australoid) proves that man evolved from several different hominid ancestors. Today, however, we know that the human race is indeed a single genetic "kind," with a built-in capacity for minor variations (e.g., in stature, skin color, hair texture, shape of eyes, nose, lips, etc.).[29] We also know that such variations can appear quite rapidly over the course of just a few generations. There is, then, nothing in our present knowledge of genetics to preclude the possibility that the human race descended from an original created pair. Indeed, given the fact that evolutionists can supply neither proof nor mechanism for the so-called "ascent of man" from (genetically) simpler creatures, the biblical model appears to be the more reasonable of the two.[30]

8. Deep Time

The currently favored version of NCE proposes that our evolving universe is about 15 billion years old. In the popular imagination, such "deep time" (i.e., lots of time, billions of years) lends credibility to NCE. People assume that given enough time anything can happen—even the emergence of a complex and orderly universe from an explosion. Now this could be true, but only if a powerful and intelligent god were involved in the ordering process. We have already seen, however, that in the godless universe of the naturalist—a universe fully subject to the Second Law of Thermodynamics and having no known mechanisms for transforming raw energy into complex material systems—such evolution (as the odds makers have unanimously declared) is impossible. Therefore, on naturalistic premises, deep time actually works *against* evolution since time plus the Second Law will always produce a *loss* of order in any physical system not preserved by "life." Deep time is no friend of NCE. Accordingly, its real relevance in the origin's debate does not lie in the fact that it supports NCE, but that, if true, it refutes recent creation. We must, therefore, revisit this subject below.

There Is Weighty Evidence against NCE

In our critique thus far we have seen that NCE violates not a few well-established natural laws and that there is little good evidence to support it. We must now conclude our evaluation by examining a few of the weightier evidences that positively speak against it. These evidences fall under two main categories, astronomical and biological.

A. Astronomical

Missing Mass: Big Bang cosmology requires a very specific amount of mass in the universe. If there were too much, the universe would have collapsed shortly after the Big Bang. If there were too little, the gravitational attraction of rapidly accelerating matter would have been insufficient to form stars and galaxies. Calculations show that there is far too little mass. Thus, in order to save the Big Bang, astronomers have rushed to the patient with a speculative *ad hoc* remedy. They postulate the existence of "dark matter"—a complex of invisible substances such as gravitons, photinos, axions, WIMPS, etc., which allegedly constitute some 90% of the mass of the universe. Unfortunately for this theory, no dark matter has actually been detected, either in the midst of galaxies or in interstellar space. Therefore, a reasonable conclusion from the actual evidence (or lack thereof) is that dark matter does not exist, the universe did not expand, and the galaxies did not evolve.[31]

Our Clumpy Universe: As we saw earlier, Big Bang cosmology argues that the CBR provides a snapshot of the universe shortly after its birth. If so, the energy in the newborn cosmos was homogeneous—evenly distributed throughout space. Looking into the heavens, however, we find that today's universe is very "clumpy." Stars and galaxies are found in clusters and other large-scale structures, all separated by great voids. In other words, matter in the universe is *not* homogeneous; it is *not* uniformly spread out in all directions. What can account for this serious discrepancy?

Big Bang theorists suggest that further explorations into deep space will yet demonstrate the universe to be homogeneous and isotropic on the largest scale. For the moment, however, the evidence is decidedly against it; and even if it were not, the CBR would still be too smooth to allow for the kind of clustering that we actually observe among the galaxies. In short, the known structure of the cosmos indicates that neither the CBR nor the cosmos arose from a Big Bang.[32]

In this connection it is interesting to note that the uneven distribution of stars and galaxies also speaks against one of the fundamental assumptions of Big Bang cosmology, the Cosmological Principle. According to this principle, the universe has no center, no edge, and no place that is "special" or atypical. Now if this were so, we would expect matter in the universe to be spread out more or less evenly so that no place *looks* special or atypical. As a matter of fact, however, some places *do* look special—and earth and the Milky Way, as we shall see, look very special. In other words, the actual structure of the cosmos speaks powerfully against the Cosmological Principle. This opens the way for a return to the more intuitive and traditional view—that the universe is a sphere in which our earth may well be at or near the center, just as the Bible appears to teach.[33]

A Cosmic Squall: Big Bang cosmology predicts that the motion of galaxies will be uniformly "outward," with all galaxies moving straight away from one another. Observations from earth show that this is not always the case. In recent years astronomers have discovered what appears to be tangential motion in millions of galaxies. That is, some galaxies—perhaps representing as much as a tenth of the universe—are not only (apparently) moving outward (as predicted) but also sideways (as not predicted). Researchers differ on the cause of this stellar "squall." But all agree on one thing: it causes big problems for the Big Bang.[34]

Galactic Evolution: It is very difficult to explain how galaxies could have arisen from a Big Bang. How could mere gravity, acting upon smoothly distributed energy, produce such diverse structures? Evolutionists respond with highly speculative answers. First, there was a brief but unimaginably rapid "inflation" during the earliest instants of the cosmos. This produced fluctuations in the energy field that served as "seeds" for the galaxies. But since these seeds were too small to account for a necessarily rapid galactic formation, the primordial gases must also have gravitated around huge clumps of "dark matter" that do not register in the CBR. However, such *ad hoc* solutions involve theoretical problems of their own, have little or no evidence to support them, and therefore remain highly controversial.

There are other questions, as well. Why are there different kinds of galaxies? How did the beautiful spiral galaxies acquire their distinctive form and rotational motion? Why do the spiral galaxies, (presumed to be) situated at vastly different distances from the earth, all look pretty much the same? That is, why do their arms all show roughly the same amount of coiling, when the further ones should show less and the nearer ones more? Why do galaxies sometimes appear in strings, at other times in clusters, and occasionally in huge walls? And why aren't these massive structures distributed uniformly throughout space? Having ruled out a divine creator, naturalists are at a loss to answer such questions. Natural laws, gleaned from the observation of existing astronomical structures, describe how they presently behave; they cannot describe how they came to be.

Of special importance here is the problem of large, complex structures appearing where they should not: at—or very near—the beginning of (the assumed) cosmic evolution. One thinks, for example, of the extraordinary 1995 photograph taken by the orbiting Hubble telescope, called Hubble Deep Field North. Aiming their telescope at a tiny patch of the northern sky in which they had formerly detected only the faintest wisps of light, astronomers were stunned to discover over 3000 mature galaxies, as well as galactic clusters. These faint, highly red-shifted objects supposedly reached their massive size and complex forms only a couple of billion years after the Big Bang. Yet on Big Bang premises, that is impossible, since a couple of billion years is much too short a time to allow for such advanced galactic evolution.

In another later study, members of the Gemini Deep Deep Survey received a similar surprise. Hoping to view scenes of the cosmos as it existed only 3-5 billion years after the Big Bang, astronomers pointed their telescope towards the so-called "Red Shift Desert," a portion of the sky thought to be among the oldest in the heavens. They too were shocked to find over 300 mature galaxies, many of them just like our own "young" Milky Way. One of the researchers involved, Dr. Karl Glazebrook, expressed his frustration this way:

> We expected to find basically zero massive galaxies beyond about 9 billion years ago, because theoretical models (*based on the Big Bang*) predict that massive galaxies form last. Instead, we found highly developed galaxies that just shouldn't have been there, but are.[35]

More than spelling trouble for the Big Bang, evidence like this offers positive support for creationist cosmology. Everywhere we look, whether near or far, we find large, mature galaxies, galactic strings, and even super-massive black holes. On Big Bang premises, many of these structures had far too little time to evolve—even if the cosmos were several times older than the presently favored 15 billion years. Thus, to use the picturesque words of one bemused observer, it looks as if someone simply "switched on the stars." Creationists couldn't have said it better themselves.

Stellar Evolution: Big Bang cosmologists theorize that the stars evolved when huge clouds of primordial gas condensed under the force of gravity and "ignited." However, astronomers have observed no such births, though they should be able to do so.[36] Furthermore, there are good reasons to believe that such births are impossible. We know, for example, that molecular "pressure" normally causes gas to disperse before gravity can act to concentrate it. Also, interstellar gas clouds typically have a high degree of angular momentum and would therefore give birth to stars rotating far more rapidly than the stars we actually observe. Finally, the magnetic fields surrounding gas clouds would powerfully resist their collapse into stars. In view of these problems many astronomers candidly admit that they do not know how stars came to be.[37]

Chemical Evolution: According to Big Bang cosmology, the elements evolved. This process involved several stages. First, raw energy from the Big Bang distilled into subatomic particles. Next, those particles distilled into clouds of hydrogen and helium gas, coexisting in a ratio of about 75% to 25%. Then the clouds of gas coalesced into a first generation of stars (Population 3 stars). In time, rising temperatures and pressures in the interior of these stars somehow produced heavier elements, such as carbon and oxygen. Finally, when these stars exploded or ejected matter into space, the residues coalesced into a second generation of stars that in turn produced within themselves the heaviest elements. This, we are told, is the origin of the Population 1 and 2 stars that we observe in galaxies today—stars whose spectra typically report high levels of "metallicity" (i.e., spectra of all but the lightest elements).

This scenario is, however, beset with theoretical and observational difficulties.

First, it is widely accepted that the initial conversion of energy into matter should have produced an equivalent amount of what is called "antimatter"—particles just like matter only with opposite charges, magnetic moments, etc. The difficulty, however, is that astronomers have been unable to locate any clusters or domains of antimatter in space. Says antimatter researcher Samuel Ting, "There have been theoretical speculations about the disappearance of antimatter, but no experimental support." Here is yet another blow to the Big Bang.[38]

Second, chemical evolution is contradicted by the Second Law of Thermodynamics. From all we can really observe, raw energy never spontaneously turns itself into matter. To the contrary, the typical effect of raw energy on matter is to destroy it, producing still more heat diffusion and still less material structure. Theoretically, energy *could* turn into matter—or simple elements into complex elements—if there were a pre-existing mechanism of some kind to effect the transformation. But naturalistic cosmology knows of no such mechanism. How, then, could raw energy from a Big Bang transform itself into more than 100 chemical elements?

There are further difficulties. We have just seen, for example, that stellar evolution cannot occur. But even if it had, we should be able to observe at least some of the Population 3 stars with zero metallicity. We do not. Similarly, on Big Bang premises, stars with very high red shifts should be primordial and therefore have few, if any, heavy elements. Observations, however, reveal the contrary. For example, light from certain remote quasars, thought to have been emitted when the universe was less than a billion years old, tells us that these objects contain more iron than our own sun! Also, certain types of stars lack helium altogether, raising further doubts about the Big Bang scenario. For these and other reasons it appears that the elements did not evolve, though they may well have been suddenly created.[39]

Planetary Evolution: According to many textbooks, our solar system evolved about 4.5 billion years ago when the force of gravity, acting upon a large cloud of swirling gas and dust, produced the sun, its several planets, and their seventy-two moons. Astronomers know, however, that this popular scenario is beset with many difficulties. For example, on these premises all the planets should spin in the same direction; in fact, three rotate backwards. All the moons should orbit their planets in the same direction; in fact, eight or more orbit backwards—and Jupiter, Saturn, Uranus, and Neptune have moons orbiting in both directions! All the moons should orbit in their planet's equatorial plane; in fact, many are in inclined orbits. All the planets should have the same basic chemical composition as the sun (98% hydrogen and helium); in fact, their compositions are radically different, not only from the sun but from each other. And so too are those of earth and its moon. The sun should have 700 times more angular momentum (i.e., high-speed spin) than the planets; in fact, the planets together have fifty times more than the sun. If the planets were formed from dust particles, we should see clear evidence of residual particles falling into the sun; in fact, we see very little. Also, observation teaches that swirling clouds of particles behave much like gas: in their gravitational interactions they do not tend to coalesce, but rather to fragment and diffuse. For these and other reasons it is hardly surprising to find astronomer Harold Jeffreys stating, "I think that all suggested accounts of the origin of the solar system are subject to serious objections. The conclusion in the present state of the subject would be that the system cannot exist."[40]

A Geo-centric Cosmos? We have seen that Big Bang cosmology presupposes the Cosmological Principle—the (unimaginable) idea that our expanding universe has no center, no edge, and, therefore, no place special. We have also seen that this presupposition does not result from direct astronomical observations but from a confessed bias against earth-centered cosmology, as well as from exotic mathematical models that seem to offer a way of avoiding it. Seekers should realize, however, that the preponderance of observational evidence decidedly favors the earth-centered view.

For example, there is no conclusive proof for the existence of other solar systems, less for other planets that support life, and still less for planets inhabited by self-conscious beings such as ourselves.

Now the Cosmological Principle predicts that all these should exist uniformly throughout the universe, but the facts, so far as we know them, show that they do not. Our solar system—and especially our earth—appear to be *very* special.

Again, there are quite a number of observations indicating that the earth, the sun, or the Milky Way may well be at the center of a spherical universe. We have already seen, for example, that quantized red shifts suggest to some that the stars are arranged in concentric shells, with earth near the center. Others have wondered if red shifts are the result of a "second-order Doppler effect"—the kind of stretching of light that occurs when a light source moves, not away from, but tangentially to, the light's recipient. In other words, galactic red shifts might mean that the universe is rotating around us! Interestingly, recently discovered patterns in the CBR point in this very direction, as do peculiarities in the way electromagnetic radiation reaches us from distant sources.[41] It is true that none of these observations are conclusive. Nevertheless, it cannot be without interest that the weight of evidence does not favor the Cosmological Principle but supports the far more intuitive view of a spherical earth or galaxy-centered universe.

Finally, we must touch here upon a very large body of evidence suggesting that our earth lies at another kind of center—the center of interest of the One who created it. Here I have in mind hundreds of phenomena indicating that earth, the solar system, and the universe itself have all been *fine-tuned* to support life on our planet. Scientists know, for example, that life could not exist if the sun were a different color, or a different mass, or closer to earth, or farther from it. The same is true of the moon: if it were only 50,000 miles closer, ocean tides would engulf nearly all earth's land mass twice a day; if slightly further, life in the stagnant seas would die. Or again, if the earth's gravity, axial tilt, rotation period, magnetic field, crustal thickness, oxygen/nitrogen ratios, and water vapor and ozone layers were only slightly different, life would perish.[42] Because, on naturalistic premises, this manifold fine-tuning is so improbable, many have concluded that there must be a Fine Tuner who has delicately structured all things for the support and enjoyment of earthly life.

In short, cosmic fine-tuning reveals earth's inhabitants as the special object of a divine creator's interest and activity. And if they lie at the center of his interest, is it not reasonable to imagine them at the center of his universe as well?

B. Biological

No Sign of Evolution: If evolution really produced the cornucopia of life forms that we see around us, we should be able to observe it at work in the present. That is, in the slice of evolutionary time we now occupy we should see multitudes of transitional life forms. These forms would display "incipient structures," structures that are not presently useful but which might (chance willing) develop into functional organs that would enhance the organism's survivability. The truth, however, is that in nature we find no such organisms. We do not find, for example, animals with useless proto-eyes, legs, wings, fins, etc. We do not find them in the present, and we do not find them in the past (i.e., in the fossil record). Wherever we look, the "evolutionary clock" appears to have stopped. There is no sign of evolution.[43, 44]

No Mechanism for Biological Evolution: Since it rejects all forms of the supernatural, NCE seeks purely natural mechanisms by which raw energy from the Big Bang is progressively transformed into simple matter, complex matter, plant and animal life, and self-conscious human beings. They are notoriously hard to find—most particularly at the stages where non-life gives rise to proto-life and where organic matter gives rise to the mystery of mind.

The proposed mechanisms for this astonishing biological transformation are *spontaneous generation* of the first cell (or cells), followed by organic evolution by means of *chance mutations* and *natural selection*. As for spontaneous generation, it is not a mechanism at all, but simply another name for a miracle without a miracle worker. As we have already seen, spontaneous generation does not and (on naturalistic premises) cannot occur. Furthermore, even if it did, chance mutations in proto-life could not have caused it to evolve, since long observation has demonstrated conclusively that mutations *never* increase the genetic endowment of an offspring, thereby introducing new biological structures. Instead, they actually *decrease* the amount of usable information—a loss that usually results in injury or death to a mutant progeny. Says Dr. Paul Grasse of the University of Paris, "The mass of evidence shows that all, or almost all, known mutations are unmistakably pathological, and the few remaining ones are highly suspect."[45]

Concerning natural selection, even evolutionists agree that it has no creative power whatsoever. That is, it adds nothing to the genetic endowment of a given species but simply eliminates its unhealthy members while preserving *variations* that are well suited to their environment. For example, in northern climes light-skinned people tend to thrive since their bodies can absorb vitamin D from the sun. Dark-skinned people, on the other hand, tend to pass from that environment since the high levels of melanin in their skin prevent absorption of vitamin D, thus exposing them to sickness. In other words, nature "selects" light-skinned people over dark-skinned people for northern climes. However, in so doing it does not create the genes that produce light skin. They were already there. Nature simply favors the survival of organisms in which those genes are dominant. Invariably, natural selection either preserves or eliminates the genetic information that is the basis of life; it cannot create it. As Dr. I. L. Cohen writes, "No one has ever produced a species by the mechanisms of natural selection. No one has ever gotten near it."[46]

We find, then, that mutation and natural selection *presuppose* the existence of complex, genetically based life. They work on what is already there and can only explain the deterioration, extinction, or preservation of what is there. They cannot explain how what is there got there in the first place. The theory of NCE, therefore, has no viable mechanism for the origin or evolution of life.[47]

The Testimony of the Experts

Our survey of evidences, which opens but a tiny window on the scientific debate about origins, nevertheless suggests that it is quite unreasonable to embrace *any* evolutionary scenario, particularly the one advanced by philosophical naturalists. Interestingly, not a few of today's scientists now agree. Indeed, because of the wealth of evidence contrary to cosmic evolution, some say the theory is in its death throes. The following quotes from respected modern scientists and commentators—none of them biblical theists—suggest that the end may be near.

- In my opinion, the observations speak a different language; they call for a different view of the universe. I believe the Big Bang theory should be replaced, because it is no longer a valid theory.[48]

  —Dr. Halton Arp

- In light of all (its) problems, it is astounding that the big-bang hypothesis is the only cosmological model that physicists have taken seriously.[49]

  —Dr. Robert Oldershaw

- A growing number of respectable scientists are defecting from the evolutionist camp... Moreover, for the most part these "experts" have abandoned Darwinism, not on the basis of religious faith or biblical persuasions, but on strictly scientific grounds, and in some instances regretfully.[50]

  —Dr. Wolfgang Smith

- I suppose that nobody will deny that it is a great misfortune if an entire branch of science becomes addicted to a false theory. But this is what has happened in Biology... I believe that one day the Darwinian myth will be ranked the greatest deceit in the history of science.[51]

  —Soren Lovtrup

- The explanatory doctrines of biological evolution do not stand up to an objective, in-depth criticism. They prove to be either in conflict with reality or else incapable of solving the major problems involved... There is no law against daydreaming, but science must not indulge in it.[52]

  —Dr. Pierre Grasse

- "Can you tell me anything you know about evolution, any one thing that is true?" I tried that question on the geology staff at the Field Museum of Natural History and the only answer I got was silence. I tried it on the members of the Evolutionary Morphology seminar in the University of Chicago, a very prestigious body of evolutionists, and all I got there was silence for a long time. Eventually one person said, "I do know one thing: it ought not to be taught in high school."[53]

—Dr. Colin Patterson, speaking in 1981 to members of the American Museum of Natural History

- Despite the fact that (we have) no convincing explanation of how random evolutionary processes could have resulted in such an ordered pattern of diversity, the idea of uniform rates of evolution is presented in the literature as if it were an empirical discovery. The hold of the evolutionary paradigm is so powerful that an idea which is more like a principle of medieval astrology than a serious twentieth-century theory has become a reality for evolutionary biologists.[54]

—Dr. Michael Denton

- I have often thought how little I should like to have to prove organic evolution in a court of law.[55]

—Errol White

- Evolution is a fairy tale for grown-ups. This theory has helped nothing in the progress of science. It is useless.[56]

—Dr. Louis Bounoure

The Attraction of NCE

But if evolution really is unreasonable, why does it *seem* reasonable to so many in our culture? The answers to this question are many. Here are a few of the most important.

First, evolution seems reasonable *because it seems analogous to the development of living beings.* All of us have observed a small, simple seed growing into a big, complex plant or tree; or a small, simple zygote growing into a big, complex human being. Such is the case, we are told, with the universe. It too has developed from an unimaginably small and simple "singularity" into the vast and complex cosmos we now inhabit—a cosmos that, just like the tree and the human being, will one day die and return to the dust.

But unfortunately for NCE, the analogy is completely false. Two hundred years ago we could not see this fact clearly; today we do—so clearly that we now view the development of individual living beings as a positive testimony *against* evolution, especially evolution of the naturalistic variety. Why is this so?

As we have already seen, a living being, *genetically considered*, always begins as a highly complex entity. The cosmos, according to NCE, did not.

A living being, *genetically considered*, does not increase in complexity as it develops. The cosmos, according to NCE, does.

A living being, at its inception, is programmed with information (i.e., its DNA) that will be used guide its development. The cosmos, according to NCE, was not.

Specific kinds of living beings, as they develop and reproduce, do not turn into other kinds of living beings. The cosmos, according to NCE, is filled with beings that continually do this very thing.

Most importantly, a living being does not develop *because* of its genetic endowment but because of the mysterious force we call "life." It is "life" that grows and animates the being *in accordance* with its genetic endowment. But the cosmos—at least according to naturalists—has no such force working in it at all.

We find, then, that the proposed analogy between organic development and cosmic evolution does not hold. At first glance it seems reasonable enough, but in the light of important biological and metaphysical truths it is finally seen to be false. Ignorance of these truths can, therefore, transform a simple perceptual error into a profoundly misplaced faith.

Second, we find that throughout our culture this (false) analogy is continually taught and reinforced by *pictures*. All of us have seen them—"artist's conceptions" of stellar evolution, planetary evolution, geological evolution, and plant and animal evolution. How well I remember from my own youth the pictures of human evolution—chains of hominids rising from stooped to upright, ugly to handsome, brutish to smart. In today's high-tech, image-driven society, such pictures have multiplied a million-fold. They appear in textbooks, posters, magazines, movies, television shows, web sites, computer games, and more. Not realizing that the pictures have no basis in observed reality, or that they are contradicted by good science, many of us simply receive them as truth. We regard them as photographs and documentaries of cosmic history. It becomes "reasonable" to believe in evolution because, after all, we have seen it for ourselves!

Seekers ought never to underestimate the power of the "icons of evolution." They can grip and dominate the imagination, especially the imagination of the young. They can shape our perception of reality. They can do our thinking for us. They can define what is "reasonable" and "unreasonable." They can quickly frame other points of view as bizarre, absurd, fantastic and even heretical, no matter how intuitive or scientifically reasonable they may be. In short, the icons can become icons indeed—religious symbols, sacred images, forming and controlling our interpretation of all reality. Let us therefore choose our pictures with great care, for it appears that false ones can hold us as tightly as we hold them, leaving little or no access for truth.

Third, cosmic evolution seems reasonable because we assume, along with the shapers of modern cosmology, that the universe is billions of years old. Most of us have grown up hearing nothing else. Like the icons of evolution, the widespread presumption of "deep time" works its way into our imaginations. It prepares the mind for evolution: what else could the cosmos do for billions of years if not evolve? It makes evolution possible: after all, anything can happen, given enough time (or so we think). The doctrine of deep time makes evolution seem necessary, true, reasonable. But is deep time really a friend of evolution, especially evolution of the naturalistic kind? The Second Law of Thermodynamics, as we have seen, says it is not. And is time really deep? Could the evidences for billions of years be as slender, equivocal, and fallacious as those for evolution itself? Careful seekers will want to know, and will work hard to find out.

Finally, cosmic evolution seems reasonable because it is believed, taught and defended by many respected authority figures in the world around us. As I sought to relate through my own story, a child's heart is prepared to trust the words of those over him—parents, teachers, pastors, journalists, authors, film-makers and more. If such authorities tell us evolution is true, we are naturally inclined to believe them. Also, the western world accords special honor to its scientists. The marvels of technology dispose us to trust them. We feel, naturally enough, that if these men understand how the world works, surely they must also understand how it came to be. Therefore, if most scientists say the world evolved, who are we non-scientists to question them? History, of course, shows that scientists can be wrong. And a little philosophy shows that "operations science" (i.e., science engaged with observable objects in the present) is fundamentally different from "origins science" (i.e., science engaged with unobservable events of the past). But most of us pay little attention to such matters. We would rather defer to those who are more intelligent, more educated, and more important than we are. It seems the reasonable thing to do—unless, perhaps, an unknown god sometimes allows the intelligentsia to stumble into foolishness; unless he expects us all to certify the truth for ourselves; unless life is a test.

There is another reason for the popularity of NCE, a reason more troubling than mere human error, gullibility, or laziness. This, however, is more appropriately discussed in another context. For now, then, we must continue with our evaluation.

Is NCE Right?

Is NCE "right"? That is, does the naturalistic creation story supply a satisfactory foundation for mankind's ethical awareness and activity? More particularly, does it adequately explain the phenomenon of conscience? Does it supply concrete moral guidance, defining for us what is good and what is evil? Does it in any way encourage goodness and restrain evil? As we are about to see, NCE does none of these things and by its very nature never can.

Our evaluation must begin with conscience. As we saw earlier, this mysterious faculty is actually a powerful "hint of a heavenly hope." When we carefully examine conscience at work, we see that man is situated in an objective moral order—an order that absolutely requires a holy god to explain its existence and operation. This order includes four basic elements: moral absolutes, moral obligation (perceived as conscience), freedom, and a law of moral cause and effect. Intuitively and inescapably, we know that certain moral absolutes exist; that we ought to align ourselves with them; that we are free to do so or not, and that, depending upon our choices, we will certainly face reward or retribution either in this life or the next. Though it is spiritual rather than physical, the moral order is no less real than the natural, and we know it. Furthermore, knowing that it is both real and spiritual, we also know that it must have been created by a god who actively rules and judges both individuals and nations through it. In short, the first four elements entail a fifth: a holy and sovereign god.

NCE, however, rules out each and every element of this order. Listen, for example, to existentialist Jean-Paul Sartre, who bluntly denied both the existence of god and moral absolutes:

> God does not exist, and we must face all of the consequences of this. The existentialist says it is very distressing that God does not exist because all possibilities of finding any values disappear. There can be no a priori good since there is no infinite and perfect consciousness to think it. So nowhere is it written that we must be honest; nowhere is it written that we must not lie, because the fact is we are on a plane where there is only us human beings. Dostoyevsky said, "If God does not exist, everything is permissible." That is the starting point of existentialism. If God does not exist, anything within or without can legitimize anything we do.[57]

But if god and moral absolutes do not exist, where do such ideas come from? What is conscience, if not the voice of god in our soul? Why do we feel free and responsible for what we do? Why do we feel that our actions are always being weighed, and that what we sow we shall surely reap?

Atheistic philosopher John Allegro replies:

> For what religious man came eventually to think of as "conscience" is simply the faculty that enabled his hominid ancestors to inhibit their programmed response to stimuli in the interests of some longer-term advantage. "Guilt" is the unease that accompanies and sometimes motivates that control, and "god" is the idealist projection of the conscience in moral terms.[58]

In other words, god, moral law, and conscience are really just biological phenomena, "designed" by life to promote the mere physical survival (or pleasure) of the living. Though it may seem otherwise, in reality our ethical ideas and intuitions have no connection whatsoever to a holy transcendent Spirit. There is no one in the cosmos (besides man) to define what is good and evil, normal and abnormal, beautiful and ugly, etc. There is no one laying down the law, no one encouraging us in our hearts to do what is right and eschew what is wrong. There is no one rewarding good or punishing evil. Indeed, there is no such thing as objective good or evil at all. Since the "good" is simply what promotes survival, the same action can be both good and evil. If lying, theft, and murder contribute to my survival (or pleasure), I have every right to call them good. Who, besides my victims, can say they are wrong?

The answer, of course, is that we all can say they are wrong, because we all *know* they are wrong. We know the objective moral order exists; we know we exist in it; and we know (or fear) that it exists in a holy and sovereign god. It is precisely for this reason that people will frequently take an action that endangers their survival. The conscience of a murderer, for example, may move him to surrender to the authorities—a decision that will obviously jeopardize both him and his progeny. Or again, whole communities may expose themselves (including their children) to imprisonment, torture, and death, rather than compromise their honor or faith. Cases like these show that conscience is (or can be) attuned to something higher and more vital than biological life itself. And it is the awareness of just such transcendent realities that inclines most people of conscience to flee the godless, amoral universe of the naturalist. Understanding that a cosmology must be right to be true, they also understand that NCE must certainly be false.[59]

In passing, I would again remind seekers that it was not until the twentieth century that mankind, in anything like significant numbers, embraced philosophical naturalism. Accordingly, the twentieth century supplies an excellent laboratory in which we may observe and evaluate its moral and societal fruits. As a matter of historical fact, these fruits appeared on the different ideological branches of NCE: Marxism, Leninism, Maoism, Nazism, Social Darwinism and Freudianism. Fair-minded observers agree that they include imperialism, racism, genocide, eugenics, coercive medical experimentation, abortion, infanticide, and euthanasia. They also include the devastating aftermaths of the sexual revolution: increased premarital and extramarital sex, divorce, homosexuality, pederasty, and pornography—all of which contribute directly to the disintegration of the physical and spiritual womb of the next generation, the nuclear family.

Is such fruit good or evil? And if it is evil, can the philosophy that spawned it be true?

Is NCE Hopeful?

Is there anything in the naturalistic creation story to communicate hope? If hope has anything to do with satisfaction of the spiritual longings underlying the questions of life, then apparently not. The reasons for this are many and clear.

Above all, NCE is hopeless because it is godless. In particular, it withholds from us a wise, good, and omnipotent creator. Such a god would himself be mankind's chief hope, and the guarantee of every other hope we might entertain as well.

Denying this creator, NCE necessarily denies the existence of a transcendent purpose for the universe, life, and man. How could such a purpose exist without a divine Someone to purpose it? Thus, in the universe of the naturalist, man has no hope of *discovering* his purpose, only of heroically *creating* one on his own. Though an occasional existentialist has bravely tried to do so, most would agree that it is a counsel of despair.

Similarly, NCE dashes our hopes of discovering how we ought to live, of attaining moral clarity for everyday life, of winning temporal and eternal rewards for resisting evil and doing good. All this, as we have seen, depends upon the existence of a holy god who, in the act of creation, establishes unchanging norms and consequences for the attitudes and actions of his free spiritual creatures. NCE denies them all.

And what of the future? There is no hope here, either, says the naturalist, since there is no supernatural soul to survive death, and no god or Heaven to whom it might return. Similarly, there is no hope for the cosmos as a whole. Since there never was an original paradise, obviously there is no hope that the cosmos will revert to it.

And since there is no god, obviously there is no hope of his creating one for us up ahead.

What hope, then, is left? The answer is clear. Having ruled out the spiritual, the naturalistic creation story shuts us up to the material. We are left by ourselves to manage ourselves as best we can amidst the mindless, purposeless engines of evolution—an explosion of matter, chance aggregations of matter, spontaneous generation of living matter, random mutation of matter, natural selection of matter, etc. How these lesser gods defied the terrible inevitability of the Second Law is beyond us. Perhaps for a while they will continue to do so. And perhaps, if we manage it well, evolution will even grant us a brief season of outstanding health and prosperity. In the end, however, the Second Law must prevail. All order, beauty, life, and consciousness must be extinguished.

Seeking to accommodate modern man to such a vision, Bertrand Russell once wrote:

> That man is the product of causes which had no prevision of the end they were achieving; that his origin, his growth, his hopes and fears, his loves and his beliefs are but the outcome of accidental collocations of atoms; that no fire, no heroism, no intensity of thought and feeling can preserve an individual life beyond the grave; that all the labors of the ages, all the devotion, all the inspiration, all the noonday brightness of human genius are destined to extinction in the vast death of the solar system, and that the whole temple of man's achievement must inevitably be buried beneath the debris of a universe in ruins—all these things, if not quite beyond dispute, are yet so nearly certain that no philosophy which rejects them can hope to stand. Only within the scaffolding of these truths, on the firm foundation of unyielding despair, can the soul's salvation henceforth safely be built.[60]

In a ghastly universe like Mr. Russell's, it is clear that poor humans have no choice but to rein in their hopes. But will they? Can they? Yes, a few stoic souls may try, but certainly not those who have come to see life as a test. For these have learned to listen to their hopes, and to hear in them the voice of an unknown god who promises that the hopes can indeed be fulfilled. Seeing, then, that the naturalistic creation story only destroys hope, they conclude that it must be false and turn away. Though uncertain of where to look next, they feel sure that somewhere out there, waiting for the earnest seeker's eye, lies a trustworthy revelation of the beginning—a revelation that will confirm their best hopes for a happy ending to the story of the universe, life, and man.

In hope they press on to find it.

Conclusion

It has been provocatively argued that in any given culture those who define the beginning are, *de facto*, that culture's true high priests. If this is so, then for a long season the high priests of Western culture have been cosmic evolutionists—usually of the naturalistic variety. Our lengthy evaluation has shown, however, that there is good reason to believe these priests have not spoken to us truly. Upon close examination, NCE turns out to be highly counterintuitive, unreasonable, ethically problematic, and devoid of any real hope. Surprisingly, it turns out there are few good reasons to be an evolutionist.

More importantly, our survey of NCE has already uncovered not a little evidence favorable to the biblical cosmology. Could it be, then, that this is the most reasonable cosmology after all? Perhaps. But before investigating it more closely we must turn aside in our journey to visit yet another popular version of the beginning—the pantheistic cosmology of classic Hinduism and the modern New Age Movement.

Take a moment to catch your breath. Though brief, this will be a tough climb!

NOTES

1. My survey of the history of cosmology draws heavily upon a book written by Dr. John Byl—*God and Cosmos: A Christian View of Time, Space, and the Universe*, (Banner of Truth, 2001) pp. 15-39. Byl is an astronomer, mathematician, philosopher, and convinced Christian. His book examines every aspect of the modern cosmological debate. On such matters, it is the single most helpful volume I have found.

2. *Ibid*, pp. 27-34. Interestingly, Byl points out that Copernicus and Galileo actually stood on no firmer ground than Ptolemy when they declared the sun to be at rest at the center of the solar system. This is because all we can ever observe is relative motion. In other words, we can know that one body is moving relative to another, but, unless a creator god exists and tells us the truth, we cannot know which body is at rest and which is moving. The heliocentric model of our solar system may be more convenient, but we cannot conclude from this fact that it is true. Says Byl, "The question as to whether it is really the earth or the sun that moves cannot be answered through scientific investigation… Ultimately, it is only God who can adequately answer the question of absolute rest and absolute motion."

3. Julian Huxley, "Evolution and Genetics," chapter 8 in *What is Science?* (J. R. Newman, ed., Simon and Schuster, 1955), p. 272.

4. Cited in Henry Morris, *The Long War Against God*, (Master Books, 2000), p. 22.

5. Many theists have, of course, tried to reconcile the two views. Some argue for what is called "theistic evolution"—the idea that God used continuing evolutionary processes in the formative side of His creative work. Others defend "progressive creation"—the idea that God created the cosmos piecemeal, bringing in new kinds of creatures at discrete intervals over the course of billions of years. Still others advocate "the gap theory"—the idea that there is a vast temporal gap between God's original creation of the universe (Gen. 1:1) and the subsequent reconstruction of a world ruined by (Lucifer's) ancient sin and judgment, (Gen. 1:2f). All such theories are motivated by a desire to harmonize the biblical creation account with (key elements of) cosmic evolution. Unfortunately, the Bible plainly contradicts them at many points. Therefore, all who embrace such theories implicitly (and sometimes explicitly) deny either the truthfulness or the clarity of the Bible on creation. In so doing, they also contradict the cosmogony of Jesus and his apostles. Seekers expecting the unknown god to grant us a clear, trustworthy revelation of the beginning will not be inclined to follow them. For a brief survey and critique of the major non-traditional views see Henry and John Morris, *The Modern*

Creation Trilogy: Scripture and Creation (vol. 1), (Master Books,1996), pp. 35-64.

6. This phrase comes from the title of Michael Denton's excellent book, *Evolution: a Theory in Crisis,* (Harper and Row, 1986). Denton, a molecular biologist, is not a creationist. Like many other professional scientists, his doubts about biological evolution arise strictly from an honest interaction with the evidence, or the lack thereof.

7. A good example is Dr. Jonathan Wells, whose book *Icons of Evolution*, (Regnery, 2000) documents his progressive realization that the traditional, textbook evidences for biological evolution are either fraudulent, inconclusive, or better explained by intelligent design.

8. Proponents of the Intelligent Design Movement include Phillip Johnson, William Dembski, Michael Behe, Stephen Meyer, and Jonathan Wells. In their books and lectures, these men argue from strictly scientific evidences that cosmic evolution does not, did not, and cannot happen—and that our fantastically complex and orderly cosmos self-evidently demands an Intelligent Designer. For more information, visit the website for The Discovery Institute, www. discovery.org.

9. Among the best are (1) *www.answersingenesis.org* (2) *www.icr. org* (3) *www.creationscience.com.*

10. The phrase is Denton's.

11. Since the early 1980s, the Gallup organization has regularly polled Americans about their beliefs on origins. The results have changed little. In 1997, slightly less than 50% of Americans believed in recent, biblical creation. About 40% believed in theistic evolution. The remaining 10% affirmed naturalistic evolution. Among scientists, the percentages differed radically: 10% chose recent creation, 40% theistic evolution, and 50% naturalistic evolution. However, the growing popularity of the Intelligent Design Movement suggests that modern scientists are increasingly open to theistic perspectives.

12. We may perform this experiment upon our own selves. I know, for example, that if I am asleep or knocked unconscious, time still exists. But if it still exists, it must still exist in a person. That person

would have to be god. Therefore I, when I finally wake up, find myself once again "in" time, because I am in god, sharing (mysteriously enough) something of his own temporal awareness of himself.

13. E. P. Hubble, *The Observational Approach in Cosmology* (Clarendon, 1937), pp. 50-51.

14. As we learn from the burning log in our fireplace, the tendency of matter is to release energy while losing structural complexity and integrity. Rarely, if ever, do we observe the opposite in nature. But NCE teaches that this is precisely how the entire universe took shape. Hence, it scandalizes our intuition.

15. This is the hidden premise of all naturalistic theories about the origin of life. When, for example, Stanley Miller, back in the 1950s, succeeded in synthesizing two amino acids by sending an electric spark through a mixture of gases thought to approximate the earth's early atmosphere, a cry of triumph rose from the Darwinist camp. Many thought we were near, not only to discovering the origin of life but to creating it! But subsequent science has not vindicated their hopes, nor can it, since the naturalistic premise of their work is wrong: that "life" is simply an electro-chemical soup with no need of a supernatural touch to transform it into a living being.

16. C.S. Lewis, in his book *Miracles* (Macmillan, 1963) shrewdly observed that "... the knowledge of a thing is not one of the thing's parts." In other words, man's knowledge of nature (and, therefore, his mind itself) is supernatural. The following poem by Emily Dickinson seems to capture the exhilaration of a mind that has just made this astonishing discovery:

> Exultation is the going
> Of an inland soul to sea,
> Past the houses—past the headlands —
> Into deep Eternity.
>
> Bred as we, among the mountains,
> Can the sailor understand
> The divine intoxication
> Of the first league out from land?

17. John Byl writes, "Since the Second Law requires order at the beginning of the cosmos, some proponents of the Big Bang assert that the original singularity was indeed orderly; that it contained 'seeds' (i.e., tiny variations of energy density) leading to the present structures. But they are baffled as to where this original order came from." They should be baffled, since such cosmic DNA needs an infinite intelligence to design it and an infinitely powerful intelligence to imprint its image upon exploding matter.

18. Science writer Sydney Harris, recognizing this fundamental problem, wistfully asks, "How can the forces of biological development and the forces of physical degeneration be operating at cross-purposes? It would take, of course, a far greater mind than mine even to attempt to penetrate this riddle. I can only pose the question because it seems to me the question most worth asking and working upon with all our intellectual and scientific resources." This quotation is cited in Henry Morris, *The Modern Creation Trilogy: Science and Creation, (Vol. 2)*, (Master Books, 1997) p.137 (Hereinafter cited as *MCT*).

19. The citations in this paragraph are all taken from Joe White, *Darwin's Demise*, (Master Books, 2001), pp. 23-50.

20. See Gene Veith, "Flew the Coop," World Magazine (December 26, 2004).

21. See *Demise*, pp. 23-50. Also, Walt Brown, *In the Beginning: Compelling Evidence for Creation and the Flood* (Center for Scientific Creation, 2001) p. 17. This book, containing the fruit of years of scientific research, is a *tour de force* of creationist evidences and theorizing. I will cite it repeatedly (as ITB). It may be purchased on-line at *www.creationscience.com*. See also, Michael Behe, *Darwin's Black Box: The Biochemical Challenge to Evolution* (New York, NY: The Free Press, 1996).

22. Using the 200-inch Mt. Palomar telescope, Dr. Halton Arp photographically verified that galaxy NGC4319 and the quasar Markarian 205 are physically connected by a bridge of luminous gas filaments. Interest here centers on the fact that the two objects have drastically different red-shifts. If, as many assume, red-shifts are

indicators of recessional velocity, then according to Hubble's law the galaxy is 107 million light years away, while the quasar is 1.2 billion! But the observed physical connection shows that this is impossible. Accordingly, Arp and others now contend that red-shifts do not indicate recessional velocity. But if this is so, what evidence do we have that the universe is expanding, or that a Big Bang really occurred?

Commenting on Arp's work, astronomer William Kaufmann writes:

> If Arp is correct (about the red-shifts not being distance indicators)...he will have single-handedly shaken all modern astronomy to its very foundations...One of the pillars of modern astronomy and cosmology will come crashing down in a turmoil unparalleled since Copernicus dared to suggest that the sun, not the earth, was at the center of the solar system.

ITB, *op. cit.*, p. 25; Byl, *op. cit.*, p. 49f. See also Don DeYoung, "The Big Bang: A Reality Check," Bible-Science News, May, 1995; and Jonathan Sarfati, *Refuting Compromise*, (Master Books, 2004), p. 156.

23. Byl, *op. cit.*, p. 55f.

24. ITB, *op. cit.*, p. 10.

25. *Ibid.*, pp. 9-10; Demise, *op. cit.*, p. 77f.

26. Demise, *op. cit.*, p. 25; ITB, *op. cit.*, p. 14. See also K. Graham, ed., *Biology: God's Living Creation* (A Beka Books, 1986) p. 361.

27. See Don Batten, ed., *The Revised and Expanded Answers Book*, (Master Books, 2000), p. 127. Also, J. Sarfati, *Refuting Evolution, Vol. 1*, (Master Books, 2000), p. 80; *Demise*, p. 99ff. Also, Marvin Lubenow, *Bones of Contention*, (Master Books, 1992).

28. Scientists have calculated that four percent of the genetic information in man differs from that of an ape. This sounds small, but actually corresponds to the amount of information contained in forty 500-page books!

29. Failure to acknowledge the crucial difference between variation and evolution ensnared many 19th and 20th century evolutionists in a deadly racism. Charles Darwin pointed the way:

> The civilized races of man will almost certainly exterminate and replace the savage races throughout the world... The break between man and his nearest allies will then be wider, for it will intervene between man in a more civilized state even than the Caucasian and some ape as low as a baboon, instead of as now between the negro or Australian and the gorilla. (Charles Darwin, *The Descent of Man*).

Following his lead, many whites henceforth perceived themselves as a distinct and more highly evolved race, while viewing Asians, Africans, "Aboriginals," Jews and Gypsies as inferior races, poorly fitted for survival. Had these "thinkers" honestly reckoned with Mendel's work (today so thoroughly vindicated), they would have realized that there is, in fact, only one race, the human race; and that in this genetically unique family, all men, despite minor variations in appearance, are brothers.

30. One outstanding piece of evidence for the unity of the human race is found in the recent discovery of Mitochondrial Eve. Though most human genetic material is located in the nucleus of our cells, a small strand of DNA may also be found in another cellular component called the mitochondria. Mitochondrial DNA (mtDNA) comes only from the mother. Though different families around the world display slight variations in their mtDNA, geneticists studying it have concluded that all human beings have descended from a single female ancestor—Mitochondrial Eve. Fascinatingly, recent studies in the rate of mutation of mtDNA indicate that Mitochondrial Eve lived about 6000 years ago. See ITB, *op. cit.*, pp. 229-30. See also, Demise, *op. cit*, pp. 23-49; MCT, *op. cit.*, pp.161-202; Answers, *op. cit.*, p. 219ff.

31. ITB, *op. cit.*, pp. 25, 35, 72, 82; Byl, *op. cit*, p. 63.

32. ITB, *op. cit.*, pp. 25, 70-71; Byl, *op. cit.*, pp. 59f, 61-63.

33. Byl, *op. cit.*, pp. 68-70.

34. De Young, *op. cit.*, p. 3.

35. Echoing Dr. Glazebrook, astronomer James Trefil speaks of galactic evolution as follows:

The problem of explaining the existence of galaxies has proved to be one of the thorniest in cosmology. By all rights they shouldn't be there, yet there they sit. It's hard to convey the depth of frustration this simple fact induces in scientists. (ITB, *op. cit.*, pp. 26-7, 73-74, 232-237).

See also Andrew Rigg, "Galaxy Games," *Creation Magazine* (December-February, 2005), pp. 18-19.

36. Displaying spectacular photos of the Eagle and Horse Head nebulae, astronomers sometimes claim we are here seeing star formation in stellar nurseries. The truth, of course, is that we are simply seeing nebulae. There are good reasons for believing that these clouds of gas *cannot* collapse into stars. See Sarfati, *Refuting Compromise, op. cit.*, pp. 167-8; ITB, *op. cit.*, 26-27.

37. Byl, *op. cit.*, pp. 53-55; ITB, *op. cit.*, pp. 25-27

38. For further reading on antimatter and the Big Bang, see the article by Michael Oard, *Missing Antimatter Challenges the Big Bang Theory*, www.answersingenesis.org.

39. ITB, *op. cit*, pp. 27, 71, 74.

40. *Ibid*, pp. 21-23, 67.

41. Sarfati, *op.cit.*, pp. 156-7.

42. See the article by Donald De Young, *The Universe is Finely Tuned for Life*, @ www.answersingenesis.org.

43. MCT, *op. cit.*, pp. 25-44.

44. NCE, with its theory of incremental change by successive small mutations, invites us to look for a *continuum* of life forms with no gaps between them. In the real world, however, we find *categories* of life forms with numerous gaps between them. It is because of these largely self-evident categories (e.g., fish, birds, insects, animals, people, etc.) that we know there is order in biological nature—an order that makes the science of taxonomy possible. The doctrine of evolution by mutation and natural selection predicts a chaos of

life-forms. The doctrine of divine creation predicts a cosmos of life-forms. The real world definitely favors the latter.

45. Cited in Demise, *op. cit.*, p. 40.

46. *Ibid*, pp. 38-43; MCT, pp. 33-45.

47. Using X-rays, scientists have been able to produce mutations in fruit flies at a greatly accelerated rate. But even after simulating many millions of years of evolutionary influences, they have been unable to produce anything other than a fruit fly. (See MCT, Vol. 1, *op. cit.*, p. 44).

48. Arp's meticulous work on red shifts and his outspoken disenchantment with Big Bang cosmology brought him into disfavor with the scientific establishment in the U.S. In order to gain access to telescopes, he was forced to migrate to Europe where he continued his labors. Despite such peer pressure, other scientists have joined him in speaking up against the Big Bang. These include Hannes Alfven, Geoffrey Burbidge, Fred Hoyle, J. V. Narlikar, A. J. Kembhavi, Eric Lerner, Robert Oldershaw, Anthony Peratt, and Tony Rothman. None, to my knowledge, are biblical creationists. For a recent examination of other non-biblical cosmologies, see Halton Arp, Roy Keys, and Konrad Rudinicki, *Progress in New Cosmologies: Beyond the Big Bang*, (New York: Plenum Press, 1985). Finally, be sure to visit *www.cosmologystatement.org*.

49. Oldershaw, Robert, "The Continuing Case for Hierarchical Cosmology," *Astrophysics and Space*, vol. 92 (1983), p. 59. For secular critiques of Big Bang cosmology, see also Eric J. Lerner, *The Big Bang Never Happened* (Random House, 1991); Fred Hoyle, *A Different Approach to Cosmology* (Cambridge, 1999).

50. See Demise, *op. cit.*, pp. 47.

51. Soren Lovtrup, *Darwinism: The Refutation of a Myth* (New York: Croom Helm, 1987), p. 422.

52. Pierre Grasse, *Evolution of Living Organisms* (New York: Academic Press, 1977), pp, 202, 103.

53. Cited in P. Johnson, *Darwin on Trial*, p. 10.

54. Denton, *op. cit.*, pp. 353-354.

55. Cited in Duane Gish, *Creation Scientists Answer Their Critics* (Master Books, 1993), p. 379.

56. Cited in Demise, *op. cit.*, p. 134.

57. Jean-Paul Sartre, *Existentialism and Humanism* (Methuen Publishing Ltd., 1974).

58. John Allegro, "Divine Discontent," American Atheist, vol. 28, (September, 1986), p. 26.

59. Since notions of moral good and evil properly belong to the theistic worldview alone, it is inconsistent, but tellingly human, to hear evolutionist Arthur Falk judge NCE as fervently as he does:

> Nature makes everything in vain. After all, what is evolution? A mindless process built on evil; that's what it is... Natural selection seems smart to those who see only the surviving products; but as a design process, it is idiotic. And the raw brutality of the process is offensive... The mix of good and evil in evolution is diabolical... In the long run, all good loses out to evil... We can believe in evolution and yet not condone it. We ought to combat it.

> *The Humanist*, vol. 55, Nov./Dec. 1995, pp. 23-25.

60. Bertrand Russell, *A Free Man's Worship*, an essay found in Campbell, Gundy, and Shrodes, eds., *Patterns for Living* (Macmillan, 1947), p. 653.

CHAPTER 3

Pantheism on the Beginning

A diligent search for the beginning will soon confront seekers with an impressive historical fact: mankind has *always* been attracted to pantheistic visions of the cosmos. Broadly speaking, these have appeared in three different forms, each arising at a different stage of world history. It is important that we examine them all with some care. For if indeed naturalistic cosmology cannot satisfy mankind's hunger for spiritual truth, then it is only reasonable to expect that many modern seekers, disillusioned with naturalism, will turn once again to pantheistic alternatives. My goal in this chapter is to help them to do so with eyes opened wide.

Let us begin with what I will call *mythic pantheism*. Cosmogonies falling into this category arose just before and during the second millennium B.C. Though geographically widespread, anthropologists agree that the earliest and most important among them are the creation myths of ancient Mesopotamia (ca. 2000 BC).

The story here begins with the Sumerians, among whom there were different, but closely related, cosmological traditions. All Sumerians believed that the universe is eternal. In the beginning it existed as a watery chaos cloaked in darkness. Then, for reasons unexplained, one or more pairs of high gods emerged from the primeval waters.

These created or gave birth to other gods, who in turn begat still others, thereby filling the vast Sumerian pantheon. In some traditions, the newborn gods simply manifested themselves *as* the earth, heaven, sun, moon, planets, and stars.[1] In others, the world is seen as the residuum of an epic battle between the gods. For example, in the Babylonian *Enuma Elish*, we learn that heaven and earth were fashioned from the dead body of the goddess Tiamat; that the Tigris and the Euphrates emerged, blood-like, from her eyes; and that man was formed from clay, mixed with the blood of the slain god, Kingu.

Though conflicting as to details, it is clear that all these cosmogonies share a common metaphysic: the universe is the embodiment of a single living substance and must be worshiped as such. Very significantly, this was the outlook of nearly all ancient cultures. Indeed, the Egyptian, Phoenician, Greek, Indian, and Scandinavian cosmogonies all bear a striking family resemblance to what many consider the Mesopotamian prototype. On the surface they are polytheistic; beneath the surface they are pantheistic, or at least pre-pantheistic.[2]

As we saw earlier, from the beginning of the first millennium BC many around the world suddenly grew weary of polytheism and myth. Henceforth, these would seek to probe the mysteries of the cosmos by reason and/or mystical insight. Not a few of them were pantheists or had pantheistic leanings. Their cosmogonies appeared from ca.1000 BC to 300 AD, a period that may be thought of as the era of *ancient religious and philosophical pantheism.*

In the East, this trend began among the Hindu Brahmins, whose metaphysical and cosmogenic speculations first appeared in the *Upanishads* (ca. 800 BC). Though Gotama himself (ca. 550 BC) did not propound a cosmogony, after his passing certain of his disciples did, clearly borrowing heavily from the developing Hindu system. Since the Hindu/Buddhist cosmology is still a live option among many modern seekers, we will examine it in some detail below.

We should note in passing, however, that Eastern cosmology is also well represented in Taoism, a religion that also has enthusiastic modern admirers. Founded by the Chinese philosopher/mystic Lao Tzu (ca. 500 BC) and his disciple Chuang Tzu (ca. 350 BC), Taoism arose in response to a sterile traditional moralism that had shown itself incapable of preserving the spiritual health of Chinese society. In setting forth their cosmology, the Taoists were trying above all to awaken their countrymen to the ultimate reality and to show them a way of living in harmony both with it and one another.

We cannot pause here to describe this worldview in detail. Suffice it to say that in all essentials Taoism is quite similar to Hinduism. It posits a single ultimate reality that is both spiritual and impersonal—the Tao. For reasons unexplained, the Tao one day manifested out of itself "the ten thousand things"—the cosmos. It must be understood, however, that these "things" are not really (material) things at all. Rather, the cosmos is the totality of phenomenal worlds existing in the minds of all sentient beings.

Upon these (phenomenal) worlds the Tao imposes a certain kind of order: the "things" and events sentient beings experience will always embody a union of two opposing principles, the Yin and the Yang. In other words, the stream of conscious experience will always be dualistic, involving subject and object, inside and outside, visible and invisible, light and dark, hot and cold, sweet and sour, love and hate, true and false, good and evil, etc. Knowing this, the wise man—and the good society—will try to live at the elusive point where the opposites meet, where Yin and Yang are in proper balance and where intimacy with the Tao is most to be enjoyed.

Again, the Taoist cosmology is quite similar to the Hindu. It has not, however, remained as popular as Hinduism since it is philosophically less comprehensive. In particular, the Taoist philosophers offered suffering humanity no clear doctrine of salvation—no instruction as to how the soul may attain release from the fetters of dualistic consciousness so as to experience final absorption into the undifferentiated oneness of the *Tao*.

During this same period pantheistic cosmologies also appeared in the West. Most historians would include among them the works of Heraclitus (ca. 500 BC), Parmenides (ca. 480 BC), the Stoics (e.g., Zeno, Seneca, and Epictetus), and the Neo-platonists (Plotinus and his disciples, ca. 300 AD). Again, space does not permit us to describe these in detail. It is, however, worthwhile to note certain essential characteristics common to all or most of them.[3]

First, these cosmologies are monistic. That is, they envision a single ultimate reality, of which all things are manifestations. Sometimes this reality is seen as ineffable and must therefore be referred to simply as "the One." Other times it is identified with a single element, such as fire. In any case, the manifold world of nature is viewed as an embodiment of the single primordial substance. That substance is eternal. And because the cosmos flows from it, the cosmos also is eternal, in one form or another.

Second, these cosmologies view the primordial substance as identical with, or indwelt by, a distinctly spiritual principle. Heraclitus, for example, declares that the world is *"an ever-living fire."* The Stoics agree, not hesitating to call this fire "god." Plotinus teaches that the One (from which the manifold world springs) is inseparable from Intellect and Soul. Here is where the cosmologies get their pantheistic feel: the ultimate reality is, or is animated by, something very much like a divine spirit or mind.

Third, these cosmologies declare that it is the spiritual principle that produces the observed order of the cosmos. In Heraclitus, for example, the *Logos* (Greek for *word, reason*) bestows upon fire the orderly forms and processes we see in nature. It transforms the primordial fire into water, and the water into earth. It governs the mingling of these elements. It maintains each element in its proper measure. It holds all the opposites in a proper balance. It maintains the rhythms of nature. Thus, as one commentator has it, "Fire is not only the material substance of all things, but a guiding and controlling intelligence (*Logos*). Nature wears the physical aspect of fire and the spiritual aspect of a cosmic reason." In short, nature is god manifesting as a visible and rational cosmos.[4]

Finally, these cosmologies usually envision the history of the universe in terms of cycles. In the Stoics we find the most severe illustration of this principle. The Stoics believed that at the beginning of each cosmic cycle the divine fire manifests as the four elements, and that these in turn develop into a new universe. When fully developed, the universe immediately begins to disintegrate, eventually reverting to the primordial fire in a great conflagration. Then the cycle begins again. Since all cycles are governed by the same absolute law of cause and effect, each successive universe will be a replica of its predecessor. The universe obeys an inviolable law of eternal recurrence.

By keeping these characteristics in mind, alert seekers will notice that some modern pantheistic cosmogonies bear a striking resemblance to their ancient Greek and Roman counterparts. More on this in a moment.

The sudden triumph of Christianity in the West brought an end to speculative philosophy in general and to pantheistic cosmology in particular. When, however, Rationalist science and philosophy began to undermine confidence in biblical revelation, pantheistic systems once again began to flourish.

It was Benedictus Spinoza who, amidst much controversy, first threw down the pantheistic gauntlet, (ca. 1650). But in things cosmological the most influential early theorists were Fichte, Schelling, and Hegel, nineteenth century German philosophers who gave to modern pantheism its distinctively evolutionary cast. This tendency was, of course, powerfully reinforced by the rise of Darwinism and NCE. Very importantly, in the 20th century we see evolutionism as the guiding motif of nearly every instance of pantheistic cosmology. We will touch on a number of them below.

As this brief historical survey reveals, there are two basic forms of pantheistic cosmology. The first, best represented by Hinduism, may be called *static*. While not denying the reality of motion or limited change, this kind of cosmology insists that a cosmos retains its basic form throughout the entire course of its existence. The second—represented by men such as Heraclitus, the Stoics, Hegel, and Teilhard de Chardin—may be called *dynamic*. It sees the form of the cosmos as continually changing, whether by evolution or devolution.

Let us pause a moment to look more closely at each.

The Hindu Cosmology

First, we have the classical Hindu cosmology. Here is a powerful picture of reality that, with notable modifications, survives today not only in modern Hinduism but in most branches of Buddhism, Theosophy, and the New Age Movement.

In one sense, it is impossible to speak of "the" Hindu cosmology since India's sacred writings, with a seemingly blithe disregard for consistency, set forth a bewildering array of "creation" myths. For example, in the old Vedic literature (ca. 1500 BC) the cosmos is represented as a building constructed by a divine carpenter, the offspring of a marriage between heaven and earth, and—most prominently—the residue of a primal sacrifice, whether of the god *Visvakarman* (Maker of All) or *Purusha* (The Cosmic Man).

The influential story of Purusha merits a retelling. When the first gods and sages issued from Purusha's being, they turned against him, pinning him down and cutting up his body. This was the primordial sacrifice, the prototype of all that will and must come thereafter. As in the case of *Tiamat*, Purusha's bodily members were now suddenly transformed, turning themselves into the earth, the sky, the sun, the moon, and more gods. Here too was the origin of mankind: from Purusha's head, trunk, loins, and feet came the four great castes of men: priests, warriors, tradesmen and servants. Observe that the *Purusha* myth supplied a very comprehensive cosmogony, unveiling the origin of the gods, the world, mankind, the social order, and the ritual sacrifices necessary to maintain the proper functioning of all things.

Later Hindu creation myths featured other gods—Prajapati, Brahman, Brahma, Vishnu, and Shiva—along with their several consorts. One of the most popular has Brahman willing into existence the primal waters. A tiny seed is floating upon them. Over the course of a year the seed grows to become a shining egg. Finally, the egg splits, giving birth to Brahma, the creator god, who first makes sky and earth out of the two halves of the eggshell and then goes on to fill the newborn universe with all its creatures.

These examples teach us that Hinduism offers not one but many different versions of the origin of the cosmos; further study would show that it gives us different versions of its structure as well. But later Hindu philosophers were not much troubled by all this diversity, for they did not regard the myths as true histories of the cosmos at all. Rather, they understood them as religious and poetic attempts to express the inexpressible: the sudden appearance of the phenomenal world(s) in the mind of Brahman. In other words, the many stories were simply human attempts to tell the one story. Thus it is indeed possible to speak of a single Hindu cosmology—though, as we are about to see, it is a story very difficult to tell.

Brahman's Dream

The Hindu cosmology begins with Brahman—the ultimate reality of Hindu theology and philosophy. The Upanishads state that Brahman is "he whom speech cannot express." Nevertheless, many Hindu teachers would affirm that Brahman may be accurately described as an infinite impersonal mind or spirit. *The Bhagavad Gita*, for example, declares, "An invisible and subtle essence is the Spirit of the whole universe. That is Reality. That is Truth. Thou art That."

Prior to the beginning, Brahman dwelt in an unimaginable state of pure, blissful, undifferentiated, spiritual oneness. This was *Nirguna Brahman*, or Brahman-without-attributes. Then, for reasons unexplained, something happened: Brahman "fell." Hindus speak of this fall by declaring that a veil of illusion, or *Maya*, settled upon Brahman. The result for Brahman was the beginning of a long and troubled cosmic dream in which Brahman, so to speak, had forgotten his true identity. The result for Brahman's offspring was the appearance of the cosmos. We must not, however, think of this newborn cosmos as an objective material reality existing independently of anyone's mind. Rather, we must see it as a vast hierarchy of phenomenal worlds, worlds existing only in the minds of a hierarchy of sentient beings, sentient beings existing only in the one mind of Brahman.[5]

To understand all this better, let us look closely at five essential characteristics of the Hindu cosmos.

The first is *consciousness*. When Brahman fell, consciousness arose. That is, Brahman himself (or itself) awoke to consciousness—but only in (or as) a host of living, sentient beings.

On the one hand, these beings were, to a greater or lesser extent, conscious of themselves. They thought of themselves as individual souls, not realizing that what they took to be their soul (atman) was actually Brahman. In other words, they awoke to consciousness but were ignorant of their true spiritual identity. They did not know that Atman (soul or subjectivity) and Brahman are one.

On the other hand, the living beings were also conscious of the particular world in which they found themselves. They thought of themselves as having physical bodies and as living in a world full of physical bodies. This world seemed real enough, as though it were an orderly assemblage of material objects independently existing "out there." But again, *Maya* had done her work. They were deceived. They did not realize that their world was merely phenomenal, that it existed only as a dream in their mind, and that their mind was, in fact, the mind of Brahman.

Here, then, is the essence of the Hindu cosmology: the cosmos is actually a vast network of interrelated dreams, all of which simultaneously sprang into existence when the One fell into the Many; when Brahman fell into consciousness; when, as it were, Brahman *fell asleep* to his own true nature and eternal bliss.

Brahman's cosmic dream also involved *hierarchy*. When the living beings awoke they found themselves inhabiting one or another of a large number of different worlds or planes (*lokas*). The sentient beings on the earth plane were arranged in a hierarchy of different populations ranging from the lowliest insect all the way up to the highest saint. Humans on the earth plane were arranged in a hierarchy of castes—social groups differentiated by birth, occupation, and rank. Meanwhile, the earth itself, usually thought of as the center of the cosmic drama, lay amidst a host of still other planes. These planes range from the lowest "hells" all the way up to the highest "heavens." Various kinds of spiritual beings, each at its own stage of spiritual development, inhabit them: gods (*devas*), demons (*asuras*), hungry ghosts (*pretas*), etc. Some are pure, others impure; some are blissful, others tormented; some are wise, others ignorant. But all, knowingly or not, await reincarnation onto the earth plane from which alone (according to many traditions) ultimate release (*moksha*) from the dream of individuality is possible.[6]

Here it is important to issue a caveat. With its intricate hierarchy of heavens and hells, the Hindu cosmos may, at first glance, seem to resemble the biblical cosmos. It must be understood, however, that the two are fundamentally different. The Bible represents the physical universe, Heaven, and Hell as true places. They are all part of a single cosmos that exists separately from God, though dependently upon him. In Hinduism, however, the various planes of the cosmos have no such objective existence. They are simply states of consciousness in the mind of Brahman.

Next, we observe that Brahman's cosmic dream is characterized by *suffering*, (*dukkha*). This suffering is ultimately traceable to the *dualistic consciousness* pervading the cosmos now that Brahman has fallen from his perfect unity. Because of the fall, living beings experience all reality in terms of dichotomies: subject-object, soul-body, mind-matter, good-evil, true-false, beautiful-ugly, healthy-sick, alive-dead, pleasure-pain, joy-sorrow, etc. Like flies in a spider's web, all creatures are stuck among them. At the start of their journey they naively cling to one or the other side of the dichotomies. In time, however, (i.e., when they become spiritually awakened human beings) they begin to realize that they cannot experience one side without the other and that their entanglement in dualistic experience is the true source of all their suffering. Accordingly, they begin to ask how they might escape this terrible bondage. Finally, they realize that *they* are not the ones seeking escape: it is actually Brahman himself, struggling to cast off *Maya*, struggling to awaken completely from the dualistic consciousness that binds him to suffering and keeps him from his original oneness and bliss. With the dawning of this awareness, sentient beings have become "path-winners." Now they are on the road to salvation.

We find, then, that in Hinduism "creation" itself introduces evil, suffering, and death into the cosmos. Here creation and fall are one. It is a truth ever to be borne in mind when comparing the Hindu and biblical cosmologies.

This brings us to a fourth characteristic, namely that the cosmic dream is governed by a principle of *spiritual ascent through reincarnation (samsara)*. That is, all living beings, whether aware of it or not, are somehow being called upward to "salvation." This salvation involves *deliverance* from the illusion of finite existence (*moksha*), and *awakening* to one's true (divine) identity (*samadhi*). Salvation is the goal and end result of all lives, deaths, and rebirths (or reincarnations). The quality of each incarnation is determined by one's *karma*, the total spiritual merit or demerit that an individual has accumulated in his previous lives. Lives basely lived will result in base incarnations, perhaps even as an unworthy animal or insect. Lives nobly lived will result in elevated incarnations, perhaps as a Brahmin or (in the Buddhist tradition) a *boddhisattva* (a savior). In the end, however, the upward call will prevail in all living beings. Having accumulated enough good *karma*, they will gain a final incarnation as a human being and then, through traditional spiritual practices and good works, achieve final salvation—complete absorption into the unconscious bliss of Brahman.

We should note in passing that this principle of ascent through reincarnation is not to be confused with cosmic evolution. In classical Hinduism there are, from the beginning, a set number of forms. Living beings, depending on their *karma*, may take different forms, but the forms themselves do not change or evolve into other forms. For this reason, many modern Hindus reject cosmic evolution. They believe that when Brahman fell, he fell into the same static cosmos (or cosmic dream) that we see around us today.

Finally, we may say that the Hindu cosmos is characterized by *eternal recycling*. The idea here is that Brahman himself is subject to a law of eternal recurrence. We see this, for example, in the Hindu conception of the days, nights, years, and lives of Brahma, the creator god. In the beginning, Brahman emanates Brahma. Brahma now begins his life. A single "day of Brahma" is a *kalpa*, and lasts about 4 billion years. At the end of one such day, Brahma dissolves the cosmos into a watery chaos and sleeps. Then, after a single night (another 4 billion years), he "awakes" and recreates the cosmos once again. Sentient beings that were not liberated during the first day are re-created on the second and continue their ascent to enlightenment. As for Brahma, he continues to wake and sleep for 100 years—the equivalent of approximately 155,500 billion earth years. Since he is now at the end of his life, he himself is received back into Brahman in the centennial dissolution. But then, after a cosmic sleep of

THE HINDU/BUDDHIST COSMOS

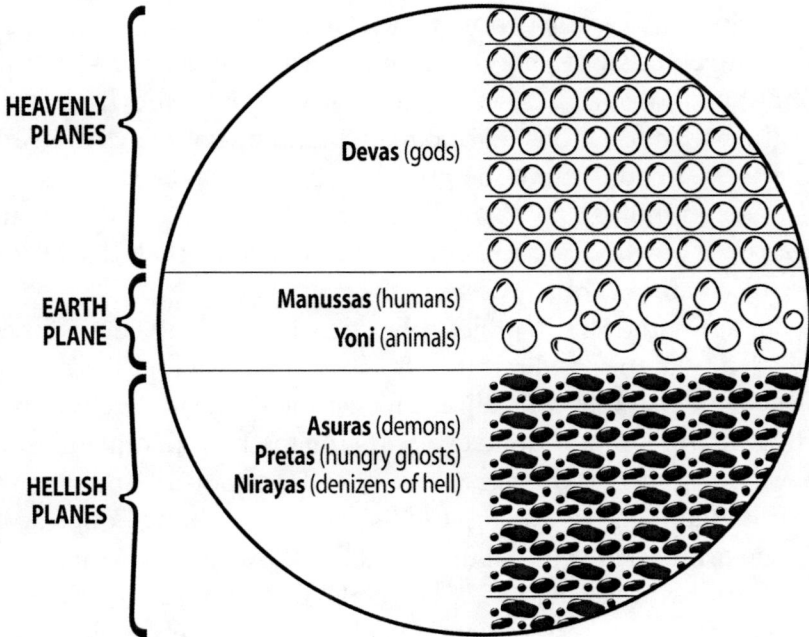

HEAVENLY PLANES

Devas (gods)

EARTH PLANE

Manussas (humans)
Yoni (animals)

HELLISH PLANES

Asuras (demons)
Pretas (hungry ghosts)
Nirayas (denizens of hell)

The Hindu scriptures supply different pictures of the cosmos. The diagram above attempts to capture their philosophical sense. The large circle represents Brahman, the ultimate spiritual reality who is one with the cosmos. At the beginning of each cosmic cycle, Brahman "falls" from his perfect oneness and is differentiated into myriad circles of consciousness, or sentient beings. In each sentient being, Brahman is dreaming a phenomenal world.

There are different kinds of sentient being. Each is arranged in a hierarchy, with its members sharing a common plane (of consciousness). Over the course of millions of lifetimes, individual sentient beings take on different forms, ascending like bubbles through the different planes. Finally, they "reach the surface," where they burst the confines of individual consciousness and dissolve into Brahman. In some traditions, salvation is from the earth plane alone — the center of the cosmic drama. Thus, Hindus and Buddhists count it a great privilege and opportunity to be incarnated as a human being.

another 100 years, a new Brahma is born. Thus the cycle begins again—and so repeats forever.[7]

In closing we should note that in the voluminous Hindu pantheon three gods hold special cosmological significance—Brahma, Vishnu, and Shiva. These constitute what is called the *Trimurti*, or the "three manifestations" of Brahman. Brahma, as we have just seen, is the creator god, Vishnu the preserver, and Shiva the completer or destroyer. In India today there are lively cults devoted to Vishnu and Shiva, both of whom are sometimes represented by their disciples as the true creator of the universe.

But again, the Hindu philosophers teach us to look beyond these three deities to the one Brahman. Some of them say that the *Trimurti* must be understood poetically as a metaphor for the threefold activity of Brahman: in each cosmic cycle, Brahma is Brahman manifesting a new (phenomenal) universe, Vishnu is Brahman sustaining it, and Shiva is Brahman destroying it—dissolving all "things" once again into the primordial spiritual unity. Others allow that the *Trimurti* are indeed true sentient beings, *devas* (gods) of very high order. Yet for all their loftiness, these too are entangled in the web *maya*, just as we poor humans. Therefore, they too are seeking salvation—release from the burdens of conscious existence and final re-absorption into the ineffable One.[8]

It is clear that the Hindu *Trimurti* reflects India's very human longing for a personal god who benevolently oversees the cosmos. So also does the polytheistic worship of her multitudes, who fear and serve millions of lesser deities. But, say her philosophers and gurus, all such gods and all their worship must pass away. In the end it is vain to look up, for there is really no one there. To find the true source—the true place of the beginning—one must look within.

Evolutionary Pantheistic Cosmology

The second kind of pantheistic cosmology reflects the efforts of modern thinkers to unite pantheistic metaphysics with evolutionism. I will call this body of thought *evolutionary pantheistic cosmology* (EPC).

Seekers will encounter two basic forms of EPC. The first might be called Western, or non-Hindu. Thinkers in this category include the early German idealistic philosophers (e.g., F. Schelling, J. G. Fichte and G. W. F. Hegel) and, more recently, Samuel Alexander, Paul Davies, Freeman Dyson, Teilhard de Chardin, Frank Tipler, and (to some extent) "process" philosophers and theologians such as A. N. Whitehead, C. Hartshorne and S. Ogden.[9,10]

To get a feel for Western EPC, let us look briefly at one of its most popular representatives, the Jesuit philosopher and paleontologist Teilhard de Chardin, (1881-1955).

Teilhard's great project was to reconcile Christianity with the newly discovered "truth" of cosmic evolution. Unlike other theologians, however, he did not pursue this goal by turning to theistic evolution. Instead he proposed a new and purportedly Christian form of evolutionary pantheism—a move that eventually elicited the censure of the Roman Catholic hierarchy.

Teilhard's thought is quite mystical and his conception of god and the beginning difficult to pin down. Brief passages in his most important book, *The Phenomenon of Man*, suggest that Teilhard's god, before the beginning, was a personal being—though certainly not the infinite tri-personal God of the Bible. If so, it was presumably at this time that he planned the forthcoming evolutionary journey of the cosmos. Then, when the beginning finally came (a beginning which Teilhard eventually identified with the Big Bang), the creator actually *became* the creation; god himself turned himself into the exploding stuff of the universe.

Echoing the views of certain ancient naturalists, Teilhard held that the resulting cosmic substance had both a material and a spiritual (or psychic) side—a without and a within (a view called *panpsychism*). In the earliest stages of evolution, the material side dominated, in which there was found only the tiniest flickers of mind or spirit. Later, however, as the original chaos evolved into a cosmos, mind began to emerge and take control. In other words, as the cosmos evolves outwardly toward greater chemical and biological complexity, it simultaneously evolves inwardly toward (fuller displays of) life, mind, self-consciousness, and—finally—god-consciousness. Thus, the ultimate purpose of the universe is for god, who *is* the universe, to awaken once again to who he is.

Teilhard called the great goal of cosmic evolution "The Omega Point" and "the Christ." In one sense, the Christ already exists: "he" is the yet future end that even now unifies and draws the universe to its appointed destiny. On the other hand, reaching the Omega point also depends upon man. This is because man, at the present stage of cosmic history, is the highest embodiment of the god who is being born. And man is free—free to manage his high destiny well, or to abort it. Therefore, through spiritually enlightened action aimed at furthering evolution and unifying the human race, man must freely cooperate with the Christ principle in order fully to realize his divine destiny. If and when he reaches the goal, man and the cosmos will have become "... a harmonized collectivity of consciousness, equivalent to a kind of super-consciousness." In this exalted state, says Teilhard, the cosmic god-man will live forever.

The second variety of EPC may be called Eastern, or modified Hindu. In this category we have the vast majority of New Agers, represented by such teachers as Fritjof Capra, Deepak Chopra, Marilyn Ferguson, David Spangler, Scott Peck, Peter Russell, John White, Ken Wilbur, Gary Zukov, and others. As a rule, these theorists attempt to synthesize elements of the classical Hindu cosmology with modern evolutionary cosmology. From Hinduism they welcome the idea that the cosmos is a manifestation of Big Mind, that there are many planes (of consciousness) in it, and that sentient beings ascend through the planes by means of reincarnation. On the other hand, they reject the Hindu idea that the cosmos is essentially static as to its form. Rather, the phenomenal world (as well as the inner world of consciousness) is continually in flux, ever evolving into new and higher forms. In the case of man, however, this evolution now appears to have become exclusively inward and spiritual. Also, New Age theorists do not teach that the goal of cosmic evolution is the absorption of the individual "soul" into Brahman's ineffable oneness. Rather, it is the deification of man who, as he approaches the Omega Point, will consciously take the reins of evolution and transform the cosmos into the world of his dreams.

Heady thoughts! Undoubtedly, they are exciting. Predictably, they are popular. But are they true—or even thinkable? We must look a little more closely to see.

Evaluating Pantheistic Cosmology

While the two varieties of pantheistic cosmology have much in common, their notable differences make it advisable to evaluate each one separately. We begin with Hindu cosmology, (HC).

Is HC Intuitive?

At first glance, the classical Hindu cosmology seems intuitive enough. We can (albeit with some difficulty) imagine a very clever Big Mind somehow "programming" himself to slip into an intricate nexus of interrelated dreams and then, from within each individual dream, to awaken to his divine nature over a long and arduous journey through many incarnations. However strange it may seem, this view definitely makes more sense than naturalism, for here, at least, there is a purposeful divine mind back of the cosmos, a mind that supplies an apparent rationale for the existence and operations of the natural, moral, and probationary orders.

And yet the more we reflect upon it, the more we realize that this is not at all what the Hindu cosmology teaches. Our natural tendency, as reflected in the language of the last paragraph, is to think of Brahman as a person—a "he" who purposes, plans, and executes his blueprint for the cosmic dream. The difficulty, of course, is that Brahman is *not* a person. Indeed, in Hinduism everything that smacks of personality—consciousness, intention, intelligence, planning, creative action, etc.—belongs to the (dualistic) realm of illusion (*maya*). If, then, Brahman truly is an "*impersonal* cosmic mind or spirit," the Hindu beginning is inconceivable and altogether counterintuitive. On the other hand, if Brahman really is a person, then Hinduism has seriously misrepresented his true nature.

But let us assume for the moment that Brahman is what he must be for this cosmology to be thinkable—a personal god. In that case we are driven to one of two cosmological options. The first is that Brahman is a kind of sado-masochist who, by intentionally slipping into a cosmic dream, subjects himself to eons of deception and suffering. This option is metaphysically conceivable but morally repugnant. It is radically counterintuitive at the moral level. The second is that there is another "god" just as powerful as Brahman (e.g., Maya), who has somehow been able to subject him to the age-long sorrows of the cosmic dream. But in this case the ultimate reality of Hinduism would not be one after all. There would be *two* ultimate realities eternally warring with one another as, for example, in the religion of Zoroaster.

We find, then, that the Hindu beginning, far from being intuitive, is actually confused and contradictory. It requires what it does not teach—a personal god. And even if it did teach a personal god, the Hindu beginning would turn him into a victim of himself or some other god more powerful than he. Such a god is many things, but he is not the holy and omnipotent unknown god revealed to us in the natural, moral, and probationary orders.

Is HC Reasonable?

Three problems make it impossible to answer this question in the affirmative.

First, there is no single Hindu cosmogony. The Hindu scriptures, as we have seen, set forth a number of different creation myths. Feeling this difficulty, the Rig Veda itself asks, "Who truly knows, who can here declare when this creation was born (and) whence it comes?" Yes, philosophically minded Hindus, looking beyond the myths to an ineffable beginning, agree that the cosmos is an emanation of Brahman. But they cannot agree about when, how, and why the (present) emanation has occurred. As opposed to its biblical counterpart, the Hindu cosmogony is vague and ill-defined. But if we cannot clearly conceive or define a cosmogony, is it reasonable to believe it?

Second, the Hindu cosmogony is without supporting evidence. Unlike the Bible, the Hindu scriptures supply no signs, no system of objective historical evidences to inspire confidence in their revelation of the beginning. Hinduism asks us to accept its version of the beginning on its own say-so. Seekers operating within the test perspective—and duly cautious about spurious revelation—will not find it reasonable to do so.

This brings us to a final difficulty—the supposed life span of the Hindu cosmos. As we have seen, the Hindu view of cosmic history is cyclical: Brahman emanates an individual cosmos and then, after about four billion years (or some such large figure), reabsorbs it back into "himself." Unfortunately, the Hindu sages supply no evidence to support this remarkable claim, while modern science is found to be continually revamping its opinions about the true age of the universe. Furthermore, it turns out there is a good deal of hard evidence—both biblical and scientific—to suggest that the cosmos is very much younger than Hinduism claims. Such evidence, which we will survey momentarily, weighs heavily against the Hindu cosmology, as also against the Big Bang hypothesis, making it unreasonable for us to believe either.

Is HC Right?

Is the Hindu cosmogony right—that is, does it win the assent of our deepest ethical intuitions? On a positive note, we may say that the Hindu idea of *karma* does indeed resonate with our moral sense. Man innately knows, as the doctrine of *karma* states, that there is a law of moral cause and effect at work in the universe. Nevertheless, other elements of the objective moral order are missing from the Hindu cosmos. Above all, Hinduism does not teach that a holy and righteous personal lawgiver created and sustains the moral order. Also, Hinduism declines to articulate a detailed moral law. Its primary ethical concern is not to define good and evil, nor to separate man from the latter to the former, but rather to transcend both through mystical experience.

Most problematic of all, however, is the metaphysical relationship of the Hindu god to moral and natural evil in the cosmos. Pantheism, as we have seen, teaches that all is one and all is god. Therefore, all that we call evil—moral depravity, calamity, sickness, suffering, and death—must be a manifestation of god. Using the vernacular of Hinduism, we must say that Brahman is both the subject and object of all evil, its perpetrator as well as its victim. The logic here is inescapable: the cosmos is both good and evil; the cosmos is a manifestation of Brahman; therefore, Brahman is both good and evil.

This conclusion deeply offends our most fundamental ethical intuitions. Intuition teaches us that the unknown god is good, that he desires the good of his creation, that he reveals the good to our conscience, and that he summons all men to it. Yes, Hinduism, with its law of *karma*, tries to express all this. But the labor is in vain. For we know innately that the moral order depends upon a moral god, a holy god, a god altogether set apart from evil. The good-and-evil god of pantheism simply does not qualify.

Pantheists, however, have an answer. They say that notions of good and evil belong to the (dualistic) world of illusion; that god is beyond all dualities, including good and evil. We cannot apply such categories to the ultimate reality. Thus Swami Muktananda declares, "Our concepts of sin and virtue... alienate us from our true Self. That which you see as impure is pure... You imagine ideas of sin and virtue through ignorance." Swami Adbhutananda agrees. "Good and evil have no absolute reality."[11]

Against all this is the voice of human intuition, assuring us that god is *only* good and not at all evil. Indeed, this awareness is so fundamental that pantheists themselves wind up using ethically charged language to describe what must surely be called the goodness of Big Mind: he (it) is *pure, adamantine, luminous, blissful, compassionate, peaceful,* etc. Now if Big Mind is good, his cosmic dream should be good as well. But in fact, the cosmos is a mix of good and evil. Therefore, the cosmic dreamer is himself a mix of good and evil.

We should note here also that it is, quite literally, nonsense to speak of a god or universe that is "beyond good and evil." The human mind, by its very constitution, seeks out and weighs the value in all it contemplates.[12] To know an object is to evaluate it, inescapably.[13] For this reason, a god "beyond good and evil" is a god beyond human conception. What our minds *can* conceive is a god that is *both* good and evil. This is the god of pantheism. Intuition assures us, however, that this is *not* the god who created and governs the objective moral order.

Since this point is so important, let me here inject a personal word.

In my four-year experiment with eastern religion, the problem of the good-and-evil god of pantheism continually haunted me. It was easy enough to scan the sky or the sea and say, "Yes, all is god." Or to peer into the throat of an orchid and say, "Yes, we are one." But I found that such affirmations caught in my own throat when evil unexpectedly intruded. I remember, for example, an afternoon in a San Francisco cafeteria when I saw a poor man fall to the floor with an epileptic seizure and nearly drown in his own vomit. That lurid scene undid months of meditation and shook my pantheistic convictions to the core. Or again, I remember browsing an issue of *Time* magazine with all due "mindfulness," only to be stung by the doleful eyes of a hungry little girl looking up into my own. Seeking a sponsor for her, the child's advocates (*World Vision*) were busy trying to alleviate evil. I was busy trying to see god's face in it. I quickly closed the magazine.

The problem with pantheism, I repeatedly discovered, was that it kept me hiding from part of reality. Part of reality is evil, and try as I would, I simply could not believe that the evil part was god. Evil, therefore, became a threat to my pantheistic faith. Had I loved the truth I would have tried to learn from evil and from my powerful intuitions about it. Because I did not, I ran from it every time we met.

Is HC Hopeful?

Does the Hindu cosmogony offer us hope? Yes and no—but mostly no.

Positively, Hinduism offers us hope in its doctrine of *samsara*—the principle of spiritual ascent through reincarnation that permeates all worlds and secretly determines the destiny of all sentient beings. In the short term, *samsara* promises us an afterlife—or rather many lives, one after another. Since, however, each of these lives is characterized by some degree of suffering, the true hope of *samsara* lies in its promise of eventual release from the wheel of life altogether, leading to final reunion with Brahman. Furthermore, the doctrine of *samsara* assures us that this hope is certain and universal: no sentient being will remain in (a consciousness of) hell forever. In time, all souls will work out their *karma*, all will awaken to their true nature, all will attain salvation—affirmations that have proved attractive to many seekers over the years.

On further reflection, however, one wonders just how hopeful this "hope" really is. Would most people really want to live—let alone suffer—through millions of lifetimes for billions of years? Could they truly take hope, not knowing for sure which will predominate in the incarnations ahead: pleasure or pain? Can they fully rejoice, knowing that even after their enlightenment the great cosmic cycle must begin again and again and again? And can this hope, such as it is, really flourish in their hearts when the truth of Hinduism is subject to so much doubt?

But even this is not all. For if we focus, not on a distant hope *beyond* this world but upon the character of our lives *within* the world, then we must confess that Hindu cosmology brings us to the brink of despair. This follows necessarily from its view of the beginning. The Hindu cosmos is *not* a purposeful creation of a holy, personal god; rather, it is the product of an inadvertent fall of an amoral, impersonal spirit. This means the world is *not* man's eternal, god-given home, but rather a fleeting and troubled dream in the mind of Brahman. If, then, man was never *intended* to live in the world, how can he possibly hope to find happiness or fulfillment in it? Even to try would be an exercise in futility, a needless entanglement in the web of Maya. In and of itself, the world is without god or god-consciousness. In and of itself, it is without meaning or purpose. In and of itself, it will never really change or improve. In and of itself, it is a realm of bondage, illusion, and suffering. Beyond this world, there is hope. In it there is none at all.[14]

Evaluating EPC

Evolutionary pantheistic cosmology is, of course, subject to nearly all the difficulties we have just examined. However, this view also involves some additional problems that merit special attention.

First, there is a (very big) metaphysical problem. Stated tersely, it is that cosmic evolution is metaphysically incompatible with pantheism.

In order to understand this, consider naturalism. Naturalism states that mind is simply a byproduct of evolving matter. However counterintuitive or irrational this assertion may be, it is at least consistent with the naturalist's metaphysical assumptions. The naturalist assumes that matter existed chronologically prior to mind, and that in special combinations with itself matter can live and produce the wonder of consciousness. Therefore, if we grant the naturalist his assumptions, his conclusion that mind is a byproduct of evolutionary processes is, metaphysically speaking, a reasonable one. Cosmic evolution and naturalism are indeed metaphysically compatible.

Similarly, cosmic evolution and theism are metaphysically compatible. This is because we can readily imagine a personal god working with created or eternally pre-existing matter so as to evolve human beings and then, at the right time, bestowing upon them the gift of consciousness.

But pantheism, by its very (metaphysical) nature, rules out cosmic evolution. This is because "matter," for the pantheist, is a byproduct of mind (Big Mind). More particularly, it is a mere phenomenon appearing in the consciousness of a sentient being. Therefore, matter cannot exist prior to the "birth" of sentient beings in whose consciousness alone it has its existence. This is exactly what the static Hindu cosmology affirms. And metaphysically speaking, the pantheist has no alternative. A sentient being may suddenly awaken to a world in the moment that Big Mind falls from its original unity. But a sentient being cannot evolve from some other substance, *since, according to pantheistic metaphysics, no such substance exists.*

In passing, we may note that Teilhard de Chardin obviously felt this difficulty and so departed from true pantheism (which states that all is divine mind or spirit) by declaring that god is a single substance that is *both* material and psychical. This exotic view, sometimes called *panpsychism*, has also been embraced by the process theologians mentioned above. However, *panpsychism* is impossible to prove scientifically—how does one observe and measure the psyche of a molecule or a cell? It is also impossible to prove spiritually, since it does not have the sanction of a credible divine revelation. Furthermore, it is liable to all the ethical problems of true pantheism since it too makes god not only the author of evil, but its embodiment as well. Panpsychism is no more reasonable—or intelligible—than true pantheistic evolution.

So again, we find after careful reflection that the doctrine of cosmic evolution is metaphysically compatible *only* with naturalism and theism, since these worldviews alone accommodate a pre-existing substance from which things (and minds) might evolve. Thus, evolutionary cosmogony cannot be grafted onto the rootstock of pantheistic metaphysics. This is why we feel such mental pain trying to imagine the beginning according to EPC. Let us learn from the pain. It is trying to teach us something important: that the much-heralded marriage of pantheism and cosmic evolution is really a shotgun wedding that will not, because it cannot, endure.[15]

Next, there is a theological problem. Fundamentally, it is that the god of EPC is so insufficiently developed at the early and middle stages of the cosmos that he (it) cannot account for the cosmos' existence, complexity, or unfolding. This god becomes personal and powerful only at the end of the cosmic journey, when at last he comes into his deity in man. But if this is so, who or what was there before the beginning to plan the cosmos (for surely a cosmos such as ours needs a plan)? Who or what manifested the initial (spiritual) singularity? Who or what oversaw its development—the progress of the phenomenal world from chaos to cosmos? And again, prior to the evolution of sentient beings, by whom or what was this progress—and all the phenomena that we call the cosmos—perceived?

In response to such questions it is, of course, possible to do as Teilhard seems to have done: posit a personal god who plans a cosmos, puts its development on a kind of spiritual automatic pilot, transforms himself (self-forgetfully) into the evolving dualistic stuff of the universe, and then, at the end of history, fully awakens to his true identity. But again, such a cosmology has no evidence to support it, and is counterintuitive, unreasonable, and morally repugnant. Lovers of science fiction may find it intriguing; lovers of truth will find it incredible.

Summing up on this point, we find that cosmic evolution is *theologically* compatible with only *one* worldview: theism. It absolutely requires an intelligent and omnipotent personal god who is able to plan, create, fashion, and direct a material cosmos to its intended goal. But EPC does not posit such a god. Therefore, its cosmology is unthinkable.

Also, it is worth remembering yet again that most of the scientific problems associated with NCE speak against EPC as well. This includes the numerous evidences cited earlier showing that evolution does not occur, did not occur, and—relative to currently functioning natural law—cannot occur. It also includes evidence (to be discussed below) indicating that the cosmos is young. Therefore, even if we grant that the inconceivable could somehow happen—that an impersonal mind could evolve a universe—the fact would remain that there is no evidence that it did, and much evidence that it did not. The evolutionary pantheist has no evidence upon which to stand. Therefore he too, like the naturalist, must take a (very great) leap of faith.

In closing, we should note that the hope of EPC differs considerably from that of classical pantheism. The Hindu hope is for an extinction of individual consciousness—the re-absorption of the self and its surrounding phenomenal universe into the undifferentiated unity of Brahman. The New Age hope is for an enlargement and collectivization of consciousness, the deification of both man and the universe as the climax of cosmic evolution. This hope also seems to involve the continuation of the phenomenal world under vastly improved conditions. Things will get better and better as man, awakening to his divine nature and power, begins to steer the evolving universe toward the world of his dreams.

If it is not humble, at least the New Age cosmology is hopeful, and far more so than the Hindu or Buddhist. Nevertheless, we have seen that there are few good reasons to believe it is true, and many good reasons to believe it is not. This is important. Hope is good, but only if it is true. Hope is true, only if it is anchored to the true future—the future that god truly has in store. And hope is powerful only if we *know* it is anchored to god's true future. The New Age hope fulfills none of these criteria. Therefore it cannot truly fill the soul with peace, nor guide it confidently through life, past death, and into its eternal home.

Conclusion

In the foregoing evaluations of naturalistic and pantheistic cosmology I have intentionally lingered long, citing many important facts and arguments relevant to the great debate about origins. My goal has been to persuade seekers that this debate is far from over, that cosmic evolution may indeed be "the great cosmogonic myth of our time," and that a paradigm shift in our thinking about origins may well be underway. Now if all this is so, seekers of cosmological truth can hardly afford to do as I did back in the seventies: to ride along carelessly on the wave of the prevailing academic consensus, making no personal effort whatsoever to ascertain the truth for oneself. Hopefully, no one who reads these pages will.

More than this, however, I have sought to demonstrate yet again the rich fruitfulness of the test perspective. As we have seen, the investigation and evaluation of different cosmologies (or competing views on any question of life) is hard work. But if we have embraced the test perspective, our load is much lightened. By it we are taught *realism:* to expect differences of opinion and a measure of difficulty in finding the truth. But by it we are also taught *optimism:* to expect that the truth *can* be found; that there is indeed a personal god; that he was present at the beginning; that he is both willing and able to give us a (much-needed) revelation of how it all came to be; and that this revelation will be understandable, intuitive, logical, supported by much good evidence, satisfying to conscience and rich with spiritual hope.

If, then, the test perspective is true, I judge that our journey so far has shown us that the naturalistic and pantheistic versions of the beginning are not.

It is time now to see if Jesus' version is.

NOTES

1. This accounts for the great importance attached to astronomy and astrology by ancient cultures. In such disciplines men were tracking nothing less than the motions of the heavenly gods and trying to determine their significance for dwellers upon the earth below.

2. The notable exception here, of course, is ancient Israel. Guided by the faith of Abraham, Isaac, Jacob, and Moses, Israel became an island of monotheistic creationism in a sea of polytheistic and pantheistic emanationism. For a fascinating attempt at tracing the development of the ancient mythic cosmogonies, see H. Morris, *The Long War Against God*, (Master Books, 2000), pp. 197-255.

3. Though not strictly pantheistic, the philosophies of Anaxagoras (ca. 450 BC) and Empedocles (ca. 450 BC) are noteworthy. Anaxagoras taught that the cosmos is the handiwork of *Nous* (Mind), an intelligent and powerful force that imposes order upon primordial "seeds" (i.e., eternal particles, lingering in the void). Similarly, Empedocles taught that the four irreducible elements (earth, air, fire, and water) are "mixed" and held in creative tension by two cosmic forces, Love and Strife. Such dualistic cosmologies, which posit a union of the material and spiritual, are usually called *panentheistic*. They argue that all things exist in, and are shaped and moved by, a god or intelligent spiritual power. The material universe is, as it were, god's body. Among the ancients, Plato is the most famous proponent of panentheistic cosmology.

4. Gordon Clark, *Ancient Philosophy*, (The Trinity Foundation, 1997), p. 37.

5. New Age author Gary Zukov gives us his own version of the pantheistic beginning, one that may serve to illumine the Hindu idea of Brahman's "fall:"

All that is can form itself into individual droplets of consciousness. Because you are part of all that is, you have literally always been, yet there was the instant when that individual energy current that is you was formed. Consider that the ocean is god. It has always been. Now reach in and grab a cup full of water. In that instant, the cup became individual, but it has always been, has it not? This is the case with your soul. There was the instant when you became a cup of energy, but it was of an immortal original Being. You have always been because what it is that you are is god, or divine intelligence; but god takes on individual forms, droplets, reducing its power to small particles of individual consciousness. (Cited in David Nobel, *Understanding the Times, op. cit.*, p. 78).

Observe here that "god" or divine intelligence does not become conscious and personal until, mysteriously enough, it "forms itself" into souls and "takes on individual forms." When the souls come into being, so too do the (phenomenal) worlds that are really within them but seem to be around them. Thus does the pantheistic universe begin.

6. One (Buddhist) version of the Hindu cosmos postulates a hierarchy of three great worlds (*lokas*): the Immaterial, the Fine Material, and the Sensuous. In the first there are four heavenly realms, populated by four different kinds of *devas* (gods). In the second there are sixteen heavenly realms, also populated by miscellaneous *devas*. The third contains eleven happy realms populated largely by *devas*, though one of them—the lowest—is the habitation of human beings. This world also contains four painful realms inhabited by demons, hungry ghosts, animals, and the denizens of hell. Importantly, there is only one realm among all thirty-one from which a sentient being can attain release from the cycle of life, death, and rebirth. That is *manussa loka*, the realm of man.

Interestingly, the author who presents this cosmology in such meticulous detail concludes by admitting, "It is pointless to debate whether these realms are real or merely fanciful metaphors describing the various mind-states we (humans) might experience in a lifetime."

The important thing, he suggests, is that the cosmology teaches us to seek enlightenment. But if we cannot trust the cosmology, are we wise to trust the path to enlightenment that it recommends?

For more detail, visit *www.accesstoinsight.org.*

7. C. Scott Littleton, ed., *Mythology*, (Duncan Baird, 2002), p. 332.

8. According to the Buddhist tradition referred to in note 6, one of the denizens of the fine material world is The Great Brahma, "… a deity whose delusion leads him to regard himself as the all-powerful, all-seeing creator of the universe." In so speaking, this cosmology expresses its preference for the impersonal "god" of pantheism and its disdain for the (deluded) personal gods of popular Hinduism and Buddhism. Visitors to Hindu and Buddhist cultures will, however, soon discover that the people themselves do not share this persuasion.

9. For a brief examination of many of these thinkers, see Byl, *op. cit.*, pp. 133-149. My own discussion of Teilhard de Chardin is highly indebted to his.

10. In strictness, process theologians must be considered *panentheists*. They teach that an impersonal force called "god" indwells eternally existing matter and transforms it into "his" (evolving) body. In man, the god-force becomes conscious and personal. From this platform it begins intentionally to guide the universe toward the fulfillment of (some of) its many possibilities. In the end, the present universe is extinguished, only to give rise to another, and so on, *ad infinitum.* (See Byl, *op. cit.*, pp. 146-149.)

11. Cited in John Ankerberg and John Weldon, *Encyclopedia of New Age Beliefs*, (Harvest House, 1996), p. 231.

12. Our minds even evaluate inorganic objects, not for moral good or evil (which attaches only to free personal agents and their acts), but for aesthetic good or evil—for beauty, for wholeness, for approximation to the mysteriously known ideal.

13. Biblical religion grounds this tendency in God himself. God evaluates all things, declaring that he is always good; that the world,

in the beginning, was good; and that the world, subsequent to man's fall, is still fundamentally good but now infected with evil. Because God created man in his image and likeness, man also evaluates all things. This is his nature, burden, and glory. To do it well is seen an essential mark of spiritual maturity, (Heb. 5:14). If, then, biblical religion is true, there is nothing in all Reality that is "beyond good and evil"—neither God nor his universe. Such a condition is a metaphysical impossibility.

14. In an utterance calculated to destroy human hope for a happy life in this world, guru Rajneesh declares, "There is no purpose in life... Life is a meaningless, fruitless effort leading nowhere... The whole of life is non-sense... You simply live; there is no purpose." (Cited in Ankerberg, *op. cit.*, p. 220).

This dreadful conclusion sets the stage for the great project of Hindu spirituality, the destruction of the human personality with a view to its final dissolution in Brahman. Having walked too many miles down that road, I would respectfully issue a solemn warning to seekers everywhere: before embracing Hindu spirituality, be absolutely certain that the cosmology upon which it is built is true. And even if you think you are, don't.

15. Interestingly, many modern Hindus recognize this problem and therefore reject attempts to unite evolutionism with their faith. A case in point is the work of M. A. Cremo and R. L. Thompson, archeologists who have written at length to show that humans have been present throughout world (and cosmic) history, just as classical Hinduism requires. See M. Cremo and R. Thompson, *Forbidden Archeology* (Bhaktivedanta Institute, 1993), pp. 797-814.

CHAPTER 4

The Teacher on the Beginning

Though Jesus of Nazareth did not teach extensively about the beginning, there can be no doubt as to his views on the subject. Along with his Jewish contemporaries, he believed and endorsed all that Moses had written in Genesis. This is particularly evident from one of his dialogues with the Pharisees concerning marriage and divorce.

> And Jesus answered and said to them, "Because of the hardness of your heart he (Moses) wrote you this precept (i.e., a law permitting divorce). But from the beginning of the creation, *'God made them male and female. For this reason a man shall leave his father and mother and be joined to his wife, and the two shall become one flesh.'* So then, they are no longer two, but one flesh. Therefore what God has joined together, let no man separate." (Mark 10:5-9)

Here we see that Jesus not only accepted the Genesis cosmogony but drew important ethical conclusions from it. In the beginning God laid down more than the physical universe. He also laid down certain norms for men and women, from which they ought not to depart. With the natural order, he created a moral order as well.

This is only one of several texts in which we find Jesus referring to the early chapters of Genesis. In speaking of the end of the world, he mentions its beginning, (Mt. 24:21). In speaking of the Sabbath, he mentions its origin in the creation week, (Mk. 2:27). In describing Satan's character, he alludes to its first display in the garden of Eden, (John 8:4). In warning of coming persecutions, he mentions their prototype in the murder of Abel by Cain, (Mt. 23:35). And in characterizing the vicissitudes of the end of the age, he recollects Noah and the flood, (Mt. 24:37-9; Luke 17:26-27). Such allusions to the beginning are hardly incidental. They give us insight not only into Jesus' view of origins but of cosmic history as well. They show us he believed that Moses had written truly about the beginning, that the ways of God and man were manifested in the beginning, and that what happened in the beginning is therefore profoundly relevant to all who live in the middle or near the end.

But what, precisely, is the biblical view of the beginning—the view that both Jesus and his disciples presupposed and referred to in so much of their teaching?[1]

This question has two possible answers, for the book of Genesis gives us both a narrow and a broad view of the origin of the world. It very much behooves us to distinguish carefully between the two.

The narrow view is found in Genesis 1-2, where we learn of the creation. These chapters follow God through his six days of creative activity, bringing the reader to *the world as it was when God took his rest.*

The broad view is found in Genesis 1-11. Here we learn not only of the *creation* but also of the *curse* that fell upon the creation as a result of Adam's sin; of a global *catastrophe* (the Flood) that completely restructured the original creation; and of a global *confusion* (at the tower of Babel) that gave rise to the diverse nations of the human family. These chapters follow God (and man) through the first 1500 or so years of cosmic history, bringing the reader to *the world as we now know it.*

If we hope fully to understand biblical cosmogony, it is vital to keep this important distinction in mind. In the beginning *week* of cosmic history God created and rested: a good beginning. In the beginning *years* of cosmic history God first created, then cursed, catastrophically judged, confused, and dispersed. Thereafter, the divine judgments ceased and the cosmos was largely fixed in its present form: all told, a bad beginning. And so, in his dealings with Abraham, God set in motion his redemptive plan. The time had come at last for a new beginning to begin (Genesis 12).

If, then, we were to ask Jesus and his contemporaries to describe "the beginning," they would likely respond, "Do you mean the good beginning or the bad, the narrow or the broad?" Again, we must understand and ever keep in mind the distinction between the two. If we don't, we cannot hope to see the cosmos as Jesus saw it. If we do, we will understand not only his view of the world, but what he hoped to accomplish by coming into it: to remove the bad, to revert to the good, and to realize all that God originally intended in the beginning.

We turn, then, to Jesus' understanding of the good beginning. To a very large extent this was shaped by Old Testament revelation, especially as that is found in Genesis 1 and 2. Here, then, is where we will resume our journey. We may summarize this aspect of OT cosmogony as follows: *In the (good) beginning, God created the heavens, the earth, the seas and all that is in them; he did so in a definite sequence, with a definite structure, and for a definite purpose; he did so in six literal days, after which he saw that all he had made was very good, rested from his creative work, and sanctified the seventh day.*

Since there is a wealth of cosmological meaning buried in this short definition, let us pause to mine each phrase just a little.

In the beginning:

As used in Genesis 1:1, this phrase appears to have a double meaning. On the one hand, it refers to "the primordial creation"— God's first creative act in which he brought into being the heavens (vacant space) and the earth (matter, submerged in the primal waters). On the other hand, it seems also to stand as a heading for the entire creation story of Genesis 1-2. In this sense it refers to our focus in this survey: "the good beginning"—that brief, six-day epoch in the history of the cosmos during which God brought into being, formed, and filled the universe for man and other living creatures.

Observe from this phrase that the biblical universe had a definite beginning. Unlike God, who exists "from everlasting to everlasting," it is not eternal, (Psalm 90:2). The kind of beginning it had is explained in the remainder of the creation story.[2]

God:

Here is the agent of creation—the infinite, personal God of the OT. In Genesis 1, he is God (Heb. *Elohim*), the powerful and majestic creator and sustainer of the universe. In Genesis 2:4ff, he is "the LORD God" (Heb., *Yahweh Elohim*), the One who, having created the world, now enters into a personal relationship and solemn agreement (covenant) with the man who is to rule it (along with his wife as helper).[3] And as we shall see momentarily, he is also the triune God, fully revealed only by Jesus and his apostles but hinted at even here in Genesis 1, where the agent of creation is three-fold: God, the Word of God, and the Spirit of God.

Created:

This word (Heb., *bara*) describes the character of God's action in the beginning. From Genesis we learn that it is essentially two-fold. On the one hand, God creates by drawing his creatures into existence by word and deed. Here we think especially of the divine fiats as, for example, when God said, "Let there be light," and there was light, (Gen. 1:3, 14). On the other hand, God also creates by forming or fashioning that which he has previously brought into being. Of special interest here is the creation of the man and the woman: the man was formed out of the dust of the ground (Gen. 2:7) and the woman was formed out of a rib extracted from the man (Gen. 2:22). As the case of the woman reveals, it is not always easy to distinguish God's bringing into being from his fashioning. But this much is sure: the biblical cosmogony is altogether unique in world religion and philosophy. As opposed to pantheistic views, it teaches that the physical universe is objectively real, external to God's being, a true creation and not an emanation or spiritual manifestation. As opposed to the naturalistic, it teaches that it had a *true* beginning: first, the universe was not, then—as the psalmist sang—"The LORD spoke, and it came to be; He commanded, and it stood firm," (Psalm 33:9, NIV).

The heavens, the earth, the seas, and all that is in them, (Exodus 20:11):

Here are the objects of creation—the cosmos as a whole. Broadly, Genesis teaches that God first created three environments—the heavens, the seas, and the earth—and then bountifully filled the three spheres with light and life. The heavens, or space, include a near heaven (i.e., the atmospheric heaven, the sky) and a far (i.e., the stellar heaven, outer space).[4] The near is bounded by waters and contains birds winging across its face. The far contains light, darkness, sun, moon, and stars. The seas contain fish and giant sea creatures. The earth contains vegetation, cattle, creeping things, beasts of the ground, and man. Israel's singers marveled at all this richness. "O LORD, how manifold are thy works! In wisdom thou hast made them all: the earth is full of thy riches", (Psalm 104:24, KJV). The fullness of the universe is testimony to the fullness of God's wisdom, power, and goodness to all.

In a definite sequence:

The biblical beginning is an orderly event, suffused with purpose and rationality. First, there is the primordial *creation* of the universe: the earth suddenly appears, suspended in vacant space, covered by water and entirely enveloped in darkness. This universe is not chaotic, but it is "formless and empty"—and therefore waiting to be formed and filled. The Spirit of God, hovering over the waters like a mother eagle above her nest, is poised to do this very thing (Gen. 1:1; Deut. 32:11, Isaiah 31:5).

Next, there is the *forming* of the universe. This occurs during the first three days of creation, as God prepares four separate environments for their respective inhabitants.

On day one, after the primordial cosmos is brought into being, God creates a mysterious, ambient light in a portion of otherwise empty space. In so doing, he separates the light from darkness so that the cycle of day and night may begin—perhaps as a result of the earth's rotation in and out of the light. The outermost heavens are now ready for the sun, moon, and stars, whose presence in the heavens will be for light upon the earth and for signs, seasons, days, months, and years (Gen. 1:1-7, 14-19).

On day two God forms another heavenly environment, separating "the waters" so as to create an "expanse" (Heb., *raqia*) between the waters that are above it and the waters that are beneath. Though this text is amenable to different interpretations, most commentators find here a reference to the creation of the atmospheric heaven—the sky—which will soon become a habitat for the birds that fly across its face, (1:20f). On this view, the waters below the expanse are the seas, and the waters above it are clouds or possibly a canopy of water vapor (Gen. 1:6-8, NIV).[5]

On day three, two more environments attain their final form as the dry land emerges from the deep and the seas fill their basins. The seas are now ready for fish; the dry lands, laden with edible vegetation, are now ready for man and the animals (Gen. 1:9-13; 2 Peter 3:5, Psalm 104:7-9).

Finally, there is the *filling* of the universe. This takes place during the last three days of creation when God sets the sun, moon, stars, fish, great sea creatures, birds, insects, land animals, and man all in their proper abodes (Gen. 1:14-31).

Note carefully that the sequence of creation evinces something important about the purpose of the universe: it is designed to be *a home for living things, and especially for man.* The prophet Isaiah set it down this way:

> For thus says the LORD,
> Who created the heavens,
> Who is God,
> Who formed the earth and made it,
> Who did not create it to be empty,
> Who formed it to be inhabited:
> "I am the LORD, and there is no other."
>
> (Isaiah 45:18)

With a definite structure:

The biblical universe, fresh from the creator's hand, was highly structured both physically and spiritually. Acting in accordance with a pre-existing plan, God impressed specific forms, functions, and relationships upon all things. In six days he brought into being "a

THE SIX DAYS OF CREATION	
Days of Forming (Habitats)	Days of Filling (Inhabitants)
Day 1 The Heavens Light & Darkness	**Day 4** The Luminaries
Day 2 The Seas The Sky	**Day 5** Fish Birds
Day 3 Dry Land Vegetation	**Day 6** Land Animals Insects Man

fixed order," after which he began to preserve, animate, and direct that order to its appointed ends (Jeremiah 31:35-36, Psalm 148:1-6).

In the creation story, examples of God-given structure abound:

The universe itself is (geocentrically) structured. This is especially clear from the primordial creation, where we see the formless earth not only at the center of empty space but also at the center of God's interest and creative activity (Gen. 1:1-2). On the fourth day the geocentricity of the cosmos will be further underscored, when God fills the heavens with "lights" that, to all appearances, revolve around the earth and exist to serve those who dwell upon it (Gen. 1:14-19).

The earth is structured. It is comprised of two main environments, the seas and the dry land. The two are separated by fixed boundaries (Job 38:8-11), and each is occupied by inhabitants specifically prepared for it (Gen. 1:9-13).

All physical things are structured. Sun, moon, and stars; seas and dry land; trees and vegetation; fish, birds, insects, animals, and men—each has its own unchanging structure direct from the creator's hand. The universe is a "fixed order," in part because all things have fixed forms and functions (Jeremiah 31:35, Isaiah 45:7). That this is the biblical view is seen especially from the case of living beings: in several broad categories (i.e., trees, vegetation, water-dwellers, creeping things, beasts of the field, etc.) God created "each according to their kind" (Gen. 1:11, 21, 24). In other words, all living beings, by creation, received from God definite physical and behavioral structures—structures that cannot fundamentally change since it is also ordained that the living beings *reproduce* "each according to its kind" (see Gen. 1:11-12, 1 Corinthians 15:39-41).

Living things are structured in a hierarchy of value. At the bottom of the hierarchy is vegetable life: grass, plants, and trees, largely serving to provide food and other necessities of life for animals and man. Next are the "living creatures" (Heb. *nephesh chayim*), distinct from vegetable life in that these have invisible souls or spirits (Heb. *nephesh, ruach*).[6] These include fish, birds, insects, and animals. And finally, ruling over all, is man. He too is a "living creature," but a supremely privileged one whose soul is uniquely cast in the image and likeness of God (Gen. 1:27-28).

Observe from all this that, biblically, biological life is seen *in conjunction* with matter but not *as a product* of matter. Biological life involves the immaterial and the supernatural. In the case of men and animals, matter is indwelt by supernatural entities: spirits. And in all cases, it is created and sustained by the Spirit of the living God (Psalm 104:30, Job 12:10). With Him is the fountain of life (Psalm 36:9). It is God alone who gives life, breath, and all things to those who live (Acts 17:25).[7]

Man is (very intricately) structured. Existing in the image of God, man is a spiritual being endowed with personality, intellect, memory, emotion, conscience, gender, rulership, and (before the fall) perfect freedom and moral rectitude. Marveling at this imprint of the divine upon a mere creature, the Psalmist exclaims that man is only "a little lower than God" (or the angels) and "fearfully and wonderfully made" (Psalm 8:5, 139:14).

Man's relationships are structured. In Genesis, we see that God created Adam and Eve in and for different kinds of relationships. He related them to himself, each other, their offspring, the animals, and the rest of the world of nature. He also revealed the privileges and responsibilities peculiar to those relationships. Here is the basis of biblical morality. What is good is what is normal: it conforms to God's norm, or design, for the relationship. Jesus' words concerning marriage illustrate this important principle. Divorce, he said, is forbidden because God created the man to cleave to his wife and to become one flesh with her. Other biblical exhortations to marital love and faithfulness, as well as prohibitions against all forms of sexual deviance, have the same creational basis. Right and wrong, in the biblical universe, depend upon the structure of things—a structure laid down by God in the beginning.[8]

In sum, the Bible teaches that at the creation the entire cosmos, both as a whole and in each of its separate parts and relations, received a fundamentally unchanging structure from the hand of God. This, by the way, was the faith of most of the founders of modern western science. Steeped in the biblical worldview, these men believed that God had created the universe according to a rational plan. That plan made their work possible and guaranteed its success. In uncovering the structures (or "laws") of nature, they were learning, as Newton declared, "to think God's thoughts after Him."[9] Naturalistic evolutionists, on the other hand, have no such basis for their scientific labors, believing that the cosmos has neither a rational creator nor any permanent structures. The two worldviews are, therefore, unalterably opposed.

For a definite purpose:

The stages and structure of creation reveal a God with a goal. Though Genesis does not exhaust the biblical revelation of God's purposes in creation, it tells us much. Broadly, we see that God created the universe—and especially the earth—for man. The biblical universe is profoundly anthropocentric.

More particularly, we see first that God intended the world to be man's *home*. It is his "proper abode," a lovingly prepared and lavishly endowed dwelling place created specially for him, (Gen. 1:29, 2:8, Psalm 115:16, Jude 6).

This world is also given to man as his *domain*, for God has specially appointed him to rule as his vice-regent over the fish, the birds, the cattle, the creeping things, and all the earth. As the psalmist prayerfully phrased it, "Thou hast put all things under his feet," (Gen. 1:26, Psalm 8).

Similarly, God purposes that the world become a kind of *workshop* in which his human children, co-laboring with their heavenly Father, fulfill a divine calling to "subdue" the earth. Sometimes referred to as "the dominion mandate," this purpose means that mankind is summoned and equipped to discover, harness, and bring forth all the hidden potentials of the natural world. The dominion mandate also involves the enlargement of the human family through reproduction, so that it can exercise a princely dominion and a loving stewardship over the entire earth (Gen. 1:26-28, 2:8, 15, Acts 1:26-28).

Finally, it appears that the cosmos was also designed as a kind of *theatre*, and this in a two-fold sense. On the one hand, it was to be a theatre in which men (and angels) could *behold the glory of God*. This means that in nature, in the marvels of his own being, and in his direct contacts with God, man would be able to grow in the knowledge of the many-faceted character of his creator. The apostle Paul affirms this purpose by declaring that in their experience of the natural world all people behold something of God's glory—his eternity, power, goodness, and other "invisible attributes" of the divine nature, (Acts 14:17, 17:25, Romans 1:20-21). In short, God created the cosmos in order to bestow upon his creatures the gift of the knowledge of himself. Interestingly, it appears that the angels too grow in their own knowledge of God's glory, especially by scrutinizing the goings-on in the earth below (Ephesians 3:8-13, 1 Peter 1:12).

On the other hand, the cosmos was also intended as a theatre in which men would *enhance the glory of God*. This does not mean, of course, that man could add anything to the perfections of the divine nature. It does mean, however, that he could bring honor to his creator in the sight of others; that he could reflect well—or ill—upon his maker, depending upon the quality of his actions before God, the angels, and other men. Not surprisingly, the Bible repeatedly exhorts us to take the high road of honoring God with our lives. Jesus, for example, commanded his followers to "Let your light so shine before men that they may see your good works and glorify your Father in Heaven," (Mt. 5:16). Similarly, the apostle Paul tersely exhorted the Corinthians, saying "Glorify God in your bodies," (1 Corinthians 6:20). Thus, one of God's high purposes in creation was to secure honor and pleasure for himself as his extended human (and angelic) family delighted in the knowledge of his glory and showed their gratitude through freely chosen acts of obedience and praise.[10]

We find then, that God had many reasons for creating the cosmos. But before any of these purposes could be fulfilled, the original pair must pass a test.

In six literal days:

The Bible is quite emphatic that God created the universe in six literal days. This foundational fact is first revealed in Genesis, a book that patently falls into the category of historical narrative. Read in its entirety, we see immediately that it is intended as a history of beginnings—whether of the universe, life, man, sin, suffering, death or God's plan of redemption. It is certainly not intended as myth or poetry.

The biblical evidence for recent creation abounds. In Genesis 1, a creation day is carefully defined as "evening and morning," the writer apparently wishing to leave no doubt as to its length. The literal view is further supported by the fact that whenever the OT uses the word "day" (Heb., *yom*) with a number (410 times), it is always a literal day. Similarly, whenever it uses the word "day" with the word "evening" or "morning" (61 times) it is again a literal day. At Sinai God confirmed the literal view when he unveiled to Israel the rationale for their Sabbath observances: "For in six days the LORD made the heavens and the earth, the sea and all that is in them, and rested on the seventh day; therefore, the LORD blessed the Sabbath (Heb., *seventh*) day and made it holy" (Exodus 20:11). The creation week was intended as the proto-type of his people's work-week, and is therefore of equal duration.

As we have seen, Jesus himself espoused recent creation, declaring that male and female were present "from the beginning of the creation" (Mark 10:6). Similarly, the apostle Paul asserted that God has revealed himself to mankind through nature "since the creation of the world" (Romans 1:20). Down through the centuries the vast majority of Christians have concurred. In modern times some interpreters, pressured by alleged scientific evidences for an old earth and universe, have tried to interpret the creation days figuratively. But even these are honest enough to admit that extra-biblical considerations alone compel them to depart from the *prima facie* sense of the text.

In short, all agree that the Bible itself unequivocally teaches a recent creation.[11]

Seekers should understand that the doctrine of a recent six-day creation is not a theological "fine point" but integral to the entire biblical cosmology and worldview.

It alone magnifies God's power—as indeed the psalmist urged Israel never to forget:

> By the word of the LORD were the heavens made,
> Their starry host by the breath of His mouth.
> He gathers the waters of the sea into jars;
> He puts the deep in storehouses.
> Let all the earth fear the LORD;
> Let all the inhabitants of the world stand in awe of Him.
> For He spoke, and it came to be;
> He commanded and it stood firm.
> Psalm 33:6-9, NIV

It alone is consistent with his manifest purpose in creation—to provide a home, a domain, and a workshop for man (Is. 45:18).

It alone supports the destiny and dignity of man. For if God's original workweek is the prototype of man's, then man's implicit destiny is to work like God and rest like God, living and serving in nature as a co-creator with him.

It alone preserves the original goodness of the universe, as well as the goodness of the One who made it. For if God's creative activity included all that the theories of an ancient universe seek to accommodate—biological trial and error, violence, bloodshed, death, and extinction—how could he, or his creation, be good?

Most importantly, it alone supports the cardinal biblical teaching about how evil, suffering, and death entered the world: through the sin of the first Adam. And this in turn supports the cardinal biblical teaching about how they will ultimately be expelled from it: through the righteousness of Christ, the last Adam (Romans 5:12ff). Here we meet very infrastructure of biblical redemption: what the first Adam lost, the last Adam gains; and what the first Adam admitted the last Adam expels. It is clear, then, that old-earth theologies, which accept the presence of natural evil in the cosmos prior to Adam, strike a mortal blow at the biblical doctrine of redemption. For if the universe is old and natural evil was present before the coming of the first Adam, what guarantees do we have that God will not include natural evil in the universe that will be re-created at the coming again of the last Adam (2 Peter 3:13, Revelation 21:1)? Yet the Bible assures us that such things will never be (1 Corinthians 15:20-28, Revelation 21:4).

In view of all this, it is hardly surprising that many theological conservatives vigorously defend the doctrine of recent creation, often at great personal cost. They believe, correctly, that the entire biblical worldview—with its unified story of cosmic creation, fall, and redemption—rises or falls with the integrity of Genesis 1-2.[12, 13]

After which He saw that all He had made was very good:

Throughout the six days of creation God saw that his work was good; on the seventh day he saw that everything he had made was *very* good, (Gen. 1:31). This recurring judgment, so manifestly exuding satisfaction, impresses upon the recipients of biblical revelation a vital cosmological truth: the world in which man now lives is not the world as it was in the beginning. Originally it was "good;" now it still is good, but also strangely mixed with evil. Originally, it knew nothing of the moral evil, guilt, sickness, death, toil, pain, and other disruptions of nature that came in with man's fall; now it does (Gen. 2-3). Accordingly, the biblical beginning fully supports a complex set of human intuitions: that the world *is good*, that it *should be better*, that *something has gone wrong*, and that *somehow it will be better again*. Similarly, this cosmogony supports our intuition that *the creator is good*, explicitly protecting him from charges that moral and natural evil sprang intentionally from his creative hand.

The doctrine of the original goodness of the creation yet again pits biblical cosmology against all forms of cosmic evolution. Cosmic evolution teaches that natural evil, in one form or another (e.g., violence, destruction, sickness, death, biological extinction, etc.) has been present in the universe from the very beginning. Moreover, in the case of theistic evolution or progressive creation, it teaches that God is the one who put it there. The Bible, on the other hand, teaches that all natural and moral evil in the universe is traceable, not to God's creation but to man's sin (Romans 5:12f, 8:18f). On this point, as on so many others, the two cosmologies are therefore altogether incompatible. It is hardly surprising, then, that every effort to reconcile biblical creation with cosmic evolution has the unintended effect of shattering the biblical worldview.

Rested from His creative work:

The divine rest does not mean that on the seventh day God stopped working in the universe (as Deism taught), only that he stopped creating (Gen. 2:1-3). In other words, he is no longer bringing new things into being (*ex nihilo*) or fashioning new things out of pre-existing material. The universe is now filled. The forms, functions, natures, and motions of things are essentially fixed. Henceforth, God no longer creates, but he is at work to sustain, animate, and direct all things to their appointed ends (Psalm 36:5f, 104)[14]

The declaration of God's creation-rest yet again puts biblical cosmology in direct opposition to all forms of evolution. The Bible states that creation was a brief once-for-all event that is now completed; evolution states that it is an ongoing process. Happily, we can easily test both views simply by looking at the world around us.

And blessed and sanctified the seventh day:

Though God created no physical objects on the seventh day, he did perform a final creative act: he blessed and sanctified the seventh day. This can only mean that he somehow impressed upon the inmost nature of his human children an inclination to set apart one day in seven to emulate him (Exodus 20:11). Accordingly, they were to rest (i.e., cease) from their work and to reflect with satisfaction upon all that God had enabled them to accomplish during the previous six days. Here, too, was a special opportunity for them to think about their creator, to ponder his plans for the future, to thank him for his many gifts—and in all of this to receive from him a special blessing. In short, by sanctifying the seventh day God instituted in man's very being a weekly rhythm of work and worship. As one thoughtful commentator has it, he "... oriented the whole created order toward the worship of God."[15]

Much later, after the fall of man, God would explicitly command his OT people Israel to observe the Sabbath (Exodus 20:8-11). Prior to the fall, however, no such command was necessary. Come the seventh day, it would only have seemed natural for Adam and Eve to join with all creation in gladly worshipping the LORD, the maker of heaven and earth (Psalm 146:1-7).

Heaven

Though Genesis 1-2 does not speak of it explicitly, other Bible passages indicate that in the beginning God brought into being a mysterious spiritual realm called Heaven. As we are about to see, forming a clear conception of its origin and nature proves challenging indeed. But let us try. We'll begin with a suggested definition of Heaven and then look more closely at each of its component parts.

Judging from the overall biblical teaching, we may say that Heaven, in the beginning, was *the place of God's continuous self-revelation to the angels, their habitation, and the point of departure for their forthcoming missions into the earth below.*

Concerning this definition, we note first of all that Heaven is a place. In particular, it is a place *above* the earth. As the psalmist wrote, "He looked down from the height of His sanctuary; from Heaven the LORD viewed the earth" (Psalm 102:19). And also, "The LORD looks down from Heaven upon the children of men, to see if there are any who understand, who seek God" (Psalm 14:2). Angels are sent down from Heaven to the earth (Dan. 10:11f, Luke 1:19, 26). Meanwhile, from the earth, prophets look up and into Heaven (1 Kings 22:19f). Indeed, one of them even ascended into Heaven bodily (2 Kings 2:1)! Jesus lifted up his eyes in prayer to God (John 11:41, 17:1). The martyr Stephen gazed up into Heaven and saw Christ at the right hand of God (Acts 8). These and many similar passages clearly represent Heaven as a place above—a place that appears to be situated in space relatively near the earth, either in the atmospheric or near astronomical heavens.

Yet despite all of this we must take care. For a careful reading of the Bible teaches us that what makes Heaven Heaven is not so much its location but the fact that God continually discloses himself there. And to understand this correctly, we must now say a few words about Heaven's first inhabitants, the angels.

The Bible teaches that in the beginning God *continuously* manifested himself only to the angels. The angels are represented as a vast host of personal spiritual beings—free and holy—whom God created early in the creation week, apparently on the first day (Psalm 148:1-5, Job 38:1-11). There were different kinds of angels (e.g., *seraphim* and *cherubim*), and also different ranks of angels, (e.g., angels and archangels). It is true that in their encounters with men, the angels appear in a given form, often human (Genesis 19:1f, Acts 2:9-11). As we shall see, however, the overall biblical testimony suggests that angels are essentially formless yet finite spiritual beings.

The angels live in Heaven—again, a place apparently situated in the physical heavens near to the earth. In Ezekiel, this place is referred to as Eden (not to be confused with the earthly Eden), the Garden of God, and the Holy Mount (Ezek. 28:11-19). Elsewhere, it is simply called Heaven (Gen. 28:12, 2 Chron. 6:21). Jesus, apparently following Ezekiel, called it Paradise, a Persian loan-word that means "a garden with a wall" (Luke 23:43, 2 Corinthians 12:4, Rev. 2:7). In Paradise, the angels behold, contemplate, and enthusiastically worship God in one or another of his visible forms (Is. 6:1f, Rev. 4, 5). As the preceding passages indicate, they perceive a Heavenly environment. In the beginning, they also observed and rejoiced as God completed his creative handiwork (Job 38:1-11). Many passages picture the angels traveling from Heaven to earth. Others teach that they were created for this very purpose: not only to enjoy the vision of God above, but to serve as his messengers to humanity on the earth below (Daniel 10:20, Luke 1:19, 26, Heb. 1:14). From these and like passages we again conclude that Heaven is, in essence, *the place of God's continuous self-revelation*—initially to the angels and later to the disembodied spirits of men as well (Heb. 12:23).

Now all this raises some difficult metaphysical questions for the biblical cosmologist. If Heaven has essentially to do with God's self-revelation, one wonders: Is Heaven a *created* world at all? That is, does it have an objective, independent existence? Is it really—as we usually imagine it to be—a world like our own, only made of a finer, more ethereal kind of matter that we humans cannot perceive because of our limited sensory apparatus? Or, on the other hand, could it be that Heaven is not a material realm after all, but rather a purely spiritual experience—a kind of extended vision existing in the minds of those men and angels to whom God grants a visible perception of invisible spiritual truths and realities?

To answer these questions it will be helpful to consider carefully Isaiah's vision of Heaven:

> In the year that King Uzziah died, I saw the Lord sitting on a throne, high and lifted up, and the train of His robe filled the temple. Above it stood seraphim. Each one had six wings: with two he covered his face, with two he covered his feet, and with two he flew. And one cried to another and said, "Holy, holy, holy, is the LORD of hosts; the whole earth is full of His glory!" (Isaiah 6:1-3)

This vision is quite similar to those received by other biblical prophets (2 Chronicles 18:18f, Ezekiel 1, Daniel 7:9-14, Revelation 4 and 5). Though they each differ in interesting and important ways, all of them include a revelation of God in a more or less human form, a throne, a temple, and angels. But does God really exist in a human form? To this fundamental question, the Bible emphatically answers in the negative, since he is essentially an infinite personal Spirit (Ex. 20:4, Deut. 4:9f). Does he then sit upon a physical throne? How could he, if he fills the whole universe (1 Kings 8:27)? Does he really live in a Heavenly temple of some kind? To Isaiah himself God said, "Heaven (i.e., the sky, space) is My throne, and earth is My footstool. Where is the house you could build for Me?" (Is. 66:1). Does he wear a robe—this One who wraps himself in light as with a garment, and stretches out the heavens like a tent (Psalm 104:2)? And what of the seraphs in heaven: do they have six wings (Is. 6:2), or four (Ezek. 1:60)? One face (Is. 6:2) or four (Ezek. 10:21)? Hands (Ezek. 10:21) or no hands, (Is. 6:2)?

Now unless the Bible is at odds with itself, there appears to be only one solution to these seeming contradictions. Heaven is not, as we often imagine it, a rarefied physical world floating like an island somewhere above the earth. Neither is it another "dimension" somehow existing apart from, but parallel to, the physical universe we humans occupy. Rather, it appears to be what may be called *a visionary world*—a world in which the presence, glory, and truth of God are revealed to angelic and human spirits under earthly imagery by means of sustained spiritual visions.

If this is so, important conclusions follow. It means that God does not really have a human form but is seen as such "in Heaven" (i.e., in heavenly visions) in order to reveal his metaphysical similarity to man. It means that he does not really sit on a throne but is seen that way in order to reveal his sovereignty over all creation. He does not really live in a temple but is seen in one to reveal his desire for the worship of his creatures in the place of his dwelling (i.e., his people, Eph. 2:22). To express all this in the words of the apostle Paul, we may say that in Heaven, as well as upon the earth, "...the invisible things of Him are clearly seen, being understood by (visions) of the things that are made" (Romans 1:20).

But does all this mean that Heaven is not a place after all? No, for we have already seen that the Bible consistently represents it as such. It does mean, however, that in light of its visionary character *Heaven's place must always be with Heaven's population.* It is a movable reality within the cosmos. On the first day it was above the earth, in the physical heavens—for that is where the angels dwelt, beholding visions of God and God's truth. Today, according to the NT, it is still above the earth, for that is where the (holy) angels and the departed spirits of the saints now dwell, also beholding visions of God, Christ, and divine truth (Heb. 12:22f).[16] Someday, however, Heaven will actually "descend" to the earth. When it does, God's ultimate purpose will be fulfilled, for that is where God plans continuously and eternally to disclose his glory to all his holy ones—much as he did among the newly created angels in the beginning (Rev. 21, 22).[17]

New Light on the Beginning

Our discussion so far has focused on the OT picture of the good beginning—a picture embraced by Jesus and his disciples. But a thorough investigation of biblical cosmogony would be incomplete if it neglected what Jesus added to this picture. What he added was something as mind-boggling as it was fundamental. He added that he himself was the creator of the cosmos!

His teaching on this crucial point is both direct and indirect, by word and by deed. Let us look at it briefly.

First, it appears that in many of his miracles Jesus sought intentionally to reveal himself to Israel as the creator.[18] In some miracles, he seems actually to have created something out of nothing. We think, for example, of how he multiplied loaves and fish to feed thousands (Matthew 14:13-21, 15:32-39, John 6:1-14). In others, he reminds us of the One who created by forming something out of what he had previously brought into being. The most familiar example here is Jesus' first miracle, performed at a wedding in Cana, when he turned water into wine (John 2:1-12). Similarly, Jesus laid his transforming touch upon a withered hand (Mt. 12:9-14), paralyzed legs (Mark 2:1-12), leprous skin (Mark 1:40-41), blind eyes (Mt. 9:27-31), and—most dramatically—dead bodies (Mark 5:35-43, cf. John 11). Of special interest is a miracle wrought upon a certain Jerusalem beggar, blind from birth (John 9:1-10). Jesus healed him by mixing spit with earth, thereafter applying the clay to the man's eyes. Reflecting upon this, his disciples may well have recalled how God formed the first man out of the dust of the ground (Gen. 2:7). In all these miracles, Jesus shows himself as creator by acting as re-creator. He restores human bodies to something of the original wholeness that God bestowed on Adam and Eve in the beginning.

In this vein, it is also worth considering the miracles in which Jesus demonstrated his power over nature. Perhaps the most dramatic of these occurred during a fierce storm on the Sea of Galilee, when he so rebuked the wind and waves that they became quiet once again (Mt. 8:23-7, Psalm 107:23-32). This display of power was not unlike that seen in the beginning when God created and shaped the elements themselves. Small wonder, then, that the disciples inquired among themselves, "Who can this be, that even the winds and sea obey Him?" The terrifying answer—that this was the creator himself—seems to have hovered, ghostlike, at edges of their shaken minds.

Turning to Jesus' teachings, we find not a little to suggest that he did indeed understand himself as the creator. In his final prayer for the disciples, he referred to himself as one who existed with the Father "before the foundation of the world" (John 17:24). To the Jewish mind, such an utterance could hardly help but raise suspicions that he was putting himself on a par with the creator. In an earlier clash with the Pharisees, Jesus was even more explicit, declaring, "Before Abraham was, I AM" (John 8:58). Here, he forthrightly takes to himself the divine Name, also suggesting that he himself was not only Abraham's seed, but his divine maker.

Finally, we have Jesus' words and deeds following his resurrection. One instance, recorded by the apostle John, is especially intriguing. Shortly after his resurrection, Jesus mysteriously appeared to his disciples in a certain house in Jerusalem. He greeted them with these words: "Peace be with you. As the Father has sent Me, I also send you." Importantly, John then relates that after so speaking, Jesus *breathed on them*, saying, "Receive the Holy Spirit." No Jew could fail to catch the symbolism here. Just as God, in the beginning, had breathed life into Adam's clay form, so now, in another kind of beginning, Jesus will soon breathe new spiritual life into his followers. Thus, in a richly symbolic act, Jesus seeks to identify himself as both the creator and re-creator of the people of God (Genesis 2:7, John 20:19-23, Acts 1:8, 2:1-4, 33).

Here we should remember also Jesus' words to John on the island of Patmos, where he appeared in a vision to his persecuted apostle. In his very first utterance, the glorified Christ identifies himself, saying, "I am the Alpha and the Omega... who is and who was and who is to come, the Almighty" (Rev. 1:8). Later, he calls himself "The Beginning and the End" and "The First and the Last" (Rev. 1:17, 2:8, 22:13). Christ's use of these exalted titles is obviously designed to comfort the harried apostle. He is reminding him that his Master is the omnipotent creator of the universe, and therefore its omnipotent consummator as well. John is to understand that he who created in the beginning may be trusted to care for his people all throughout the middle—and then to return, resurrect, and re-create in the end (Rev. 21:5).[19]

If, however, we limit ourselves to his pre-resurrection teaching, it does appear that Jesus made no explicit claim to being the creator. What might have been his reasons for this? Biblical conservatives reply by arguing that he sought, like all previous and subsequent biblical teachers, to ascribe creation preeminently to God the Father (Mt. 5:43-48, 11:25,19:6). More importantly, he also refrained because he knew that his disciples had not yet fully understood the trinitarian mystery—nor could they until the coming of the Holy Spirit (John 16:13). This view seems to be vindicated by the events of NT history. When at last the apostles did receive the Spirit on the day of Pentecost, they soon after began declaring not only the deity of their Master but also his now more fully revealed role in the beginning (see Mt.11:25-30).

That role can be distilled into the little word "through." Over and again in the NT, we find the apostles teaching that the place of the Son in the divine economy is to act as the mediator *through* whom the Father relates to the world. This applies to the creation itself. For them, the Son was the divine co-creator—the One through whom the Father brought into being the entire cosmos.

This understanding is most vividly on display in the well-known prologue to John's Gospel:

> "In the beginning was the Word, and the Word was with God, and the Word was God. He was in the beginning with God. All things were made through Him, and without Him nothing was made that has been made" (John 1:1-3).

These rich lines, so pointedly reminiscent of Genesis, not only identify Jesus as the creator but explain his (pre-incarnate) role in the beginning. When God created, says the apostle, it was the Father who devised the plan and the Son who gave the word.[20] The role of the Son was to be the executor or administrator of God's creative acts. Thus when we read over and again in Genesis 1, "Then God said," John would have us to understand that it was the Son who did the speaking. Serving as the dynamic Word of God, he was mediating the creation of the cosmos. Notably, the other apostles echo John on this crucial point. All of them believed and affirmed that *through* their divine, miracle-working Teacher, all things, whether in Heaven or on the earth, had been made (John 1:3, 1 Corinthians 8:6, 1 Timothy 2:5, Heb. 1:1, etc.).

In this connection, we must not fail to touch on a New Testament passage that throws important new light on Christ and on God's purpose in creation. It is found in Paul's letter to the Colossians. Addressing various misconceptions about God's Son, he writes as follows:

> He is the image of the invisible God, the first-born over all creation. For by Him all things were created, both in the heavens and on earth, visible and invisible, whether thrones or dominions or rulers or authorities—all things have been created through Him and for Him. And He is before all things, and in Him all things hold together (Colossians 1:15-17).

These rich words go far to flesh out the total biblical teaching on Christ and creation. Like John, Paul teaches that God created the cosmos through his divine Son. Going further, he also teaches that the Son now holds the cosmos together in its appointed structure. Going further still, he then declares something new and unexpected: that the Father not only created the cosmos *by* the Son, but also *for* the Son. Here is fresh biblical light on the purpose of God in creation. God created the cosmos not only for his own glory and the good of man, but also for the purpose of expressing, quite publicly, his love for his Son (Col. 1:13). From the beginning, says Paul, he intended to accomplish this by placing all things in the custody of the Son and under the sovereignty of the Son. The Son was to be "the first-born over all creation," even as the sons of earthly patriarchs in biblical times were to be over all that they had inherited from their fathers. In short, the creation reflects a love affair between the Father and the Son!

Interestingly, the entire drama of redemption may be cast in terms of this amazing revelation. In creation, God planned to exalt his Son as Lord over all. In the fall, the plan seemed certain of failure. But in redemption it was realized at last, and that in a manner more marvelous than if there had been no fall at all. For the moment, says Paul, the world remains largely ignorant of this mystery. But in the end all will finally realize that the Father loves the Son, that he created the universe for him, and that he has effectively delivered all things into his hands against the day when he (Christ) will return to make all things new (John 5:20, Matthew 11:27, Ephesians 1:9-10, Philippians. 2:5-11, 3:20-21).

Summary

It is time now to summarize the Teacher's view of the origin of the universe, life, and man—and of how we may certainly know the truth concerning this fundamental question of life.

Jesus of Nazareth, along with his apostles, invites seekers to turn to the Bible, believing that there alone they will find a complete and trustworthy divine revelation of the beginning. This revelation begins in Genesis, which gives a foundational account of the good and bad beginning. It is supplemented by other OT passages, and brought to completion by Jesus himself, whose cosmological teachings are recorded throughout the New Testament. God Himself, by means of the Holy Spirit, is able to show and assure every sincere seeker that this revelation of the beginning is indeed the truth, (Mt. 11:25, John 7:17, 16:13).

The revelation takes us back to a time before the creation of the world, when nothing existed except the infinite, tri-personal God, living in eternal fellowship as a Holy Family comprised of Father, Son, and Holy Spirit. In this primordial state, God the Father determined, as it were, to enlarge his family by creating a race of men and angels, along with a lavishly adorned physical abode in which they were to live. He decided to create the universe. It would be, in essence, a home, a domain, and a workshop for the most privileged member of the extended family—man. Here the Father's human children would have important and delightful work to do: under his Son—and with occasional assistance from the angels—they would co-labor with God himself in the fulfillment of his secret plans for the world.

In this great enterprise, all the parties in the relationship would be blessed and God's manifold purposes for the cosmos thereby fulfilled. God would bless man with an ever-deepening knowledge of his glory, even as man would bless God with an ever-deepening gratitude, enjoyment, and obedience toward him. Furthermore, because the Father would appoint the Son as head over all, the Son would be especially honored and the Father especially pleased.

With these and other purposes in mind, God, in the beginning, took action. Through his Son and by his Spirit he created the cosmos, calling it into being and fashioning it into a habitable world in six literal days. The first three days were days of forming, when he created three separate environments: the heavens, the seas, and the dry land. The second three days were days of filling, when he set into their respective habitats the heavenly bodies, birds, fish, insects, animals, and—with dominion over all other earthly creatures—man.

Also, apparently on the first day, he created the angels. These invisible personal spirits lived in Heaven—a visionary world within the cosmos where they enjoyed an ongoing revelation of God and divine truth. From Heaven above God would regularly send them down to minister to his children on the earth below.

This universe, fresh from the creator's hand, was a true cosmos: orderly, beautiful, and highly structured—not only physically but spiritually. It was also good, with no admixture of evil whatsoever. Though ready for development at the hands of man, it was—and was to remain—a fixed order that would never pass away. It would be an eternal dwelling-place for God and the family of man.

Having completed his creative work, God rested on the seventh day, ordaining that his human children should do likewise until their cosmic vocation was complete. Thus did God orient his creation to the worship of himself, inclining his human children regularly to reflect on their heavenly Father, thank him for his gift of life and all things, and prayerfully ponder what yet remained to be accomplished in their journey through history with the creator of the cosmos.

All this and more awaited Adam and Eve—and the children they soon were to bring into the world. But to enjoy it forever, they first must pass a test.

Here, then, in too few words, is something of Jesus' view of the good beginning. It was, for him, a stupendous cosmological reality upon which his eyes were ever fixed. Yes, he understood all too well the tragic events that quickly followed that beginning: the failed probation of Adam, the cursing of the cosmos, the flooding of the whole earth, the confusion of language, and the dispersion of the nations from Babel—everything that bequeathed to the sons of Adam the (broken) world into which he himself was born. But none of it daunted him. For he also knew that in his own day One had come who would redeem the cosmos; who would bring back and gather together a holy family for God's possession; who, in due season, would recreate the world, and who would make of it all it was meant to be in the beginning.

Jesus knew that such a One had come—and he fervently believed, taught, and demonstrated that it was none other than himself.

Evaluating Biblical Cosmology

As we begin our evaluation, let us note immediately that the Teacher's cosmogony is not only highly intuitive, but far more so than that of naturalism or pantheism. The reason is clear: the biblical cosmogony posits a personal creator god—the kind of god we see reflected in countless ways in the natural and moral orders. Naturalism and pantheism do not. They fail to satisfy because they do not build upon the only possible foundation for a sound cosmogony: a wise, holy, and all-powerful personal supreme being—just the kind of supreme being the Bible proclaims.

But let us be more specific, touching briefly on both the intuitiveness and reasonableness of some key elements in the biblical creation account.

The biblical assertion that the universe is not eternal but had a definite beginning is highly intuitive. Many things in our experience, including our own selves, have a beginning. It is only reasonable to expect the same of the cosmos. Interestingly, science now strongly confirms this very conclusion. On the one hand, there is no evidence at all for the continuous creation of matter, nor for the continuous formation of new kinds of beings. On the other, there is abundant evidence that all existing beings decay according to the Second Law of Thermodynamics. Taken together, these two facts mean that our orderly universe must somehow have sprung into existence fully formed and functioning, after which, for some reason, it began to run down and fall apart. It is, therefore, both intuitive and reasonable to agree with the Bible that the universe had a definite beginning.

It is also quite easy to believe that in the beginning a *personal* god created the universe. Innately, we expect this very thing. What's more, our expectation is only strengthened by the evidence. Everywhere we turn, whether within or without, we observe order, design, purpose, beauty, and benevolence—all marks of a decidedly personal creator and sustainer. That this creator should be revealed to us as three persons in one is indeed unexpected, but not counterintuitive. Furthermore, under the light of this revelation certain peculiarities of the natural world begin to make sense. Why, for example, should the one phenomenon of space be comprised of three dimensions—length, width, and depth? Why should time be triune, involving past, present, and future? Why does matter in the cosmos exist simultaneously in three basic forms—solid, liquid, and gas? And why is the cosmos itself—or at least our experience of it—triune, consisting of time, space, and matter? In the light of biblical revelation, such "little trinities" speak loudly, bearing witness to a triune creator who has left the imprint of his own nature upon the world.

The "dualistic" cosmos of the Bible—a cosmos that is both spiritual and physical—also appeals to our spiritual common sense. Innately, we are all dualists. We cannot long abide monism, whether materialistic or pantheistic. We know that *god* is different from matter, though related to it. And we know that *we* are different from matter, though related to it. Also, we have no difficulty accepting the idea of angels, or of a Heaven in which the angels live before the face of God. In the nature of the case, the existence of such spirit beings cannot be confirmed scientifically. But anecdotal evidence for their existence is abundant and significant. In weighing it, wise seekers will be duly cautious but not closed-minded. They know that where there is much smoke there is likely to be real fire.

For those steeped in evolutionary cosmogonies, the biblical story of a brief, six-staged creation, in which three environments are formed and filled, does indeed strike one as strange. The story itself, however, contains nothing counterintuitive. Indeed, on the assumption that God created the universe for living things—and especially for man—such a simple procedure is not only reasonable but predictable. Why would God use age-long evolutionary processes to create a home for occupants who would not appear on the scene for billions of years? Logically, an anthropocentric universe cries out for a brief, anthropocentric creation. What's more, most of us find the idea of an anthropocentric creation quite intuitive. It is only common sense to conclude from a universe so manifestly suited to man that it was actually created with man in mind. And this, interestingly enough, is precisely the conclusion of many modern scientists. Observing what they call "the anthropic principle" pervading the cosmos, they declare that "The universe seems to be tailor-made for man."[21] A brief, man-centered, six-staged creation is intuitive, logical, and evidentially confirmed. Many, therefore, find it reasonable to believe.

The same is true of the Bible's account of the many structures that God created. It is intuitive because it harmonizes so wonderfully with our everyday experience of the heavens, the earth, inanimate objects, living things, man, and man's relationships. Looking upon them, we see immediately that none of these things change in any fundamental way. Once for all they have received structures—structures manifestly based upon a specific purpose and design. Thus do they testify loudly to the kind of rational god we meet in the Bible.

The Bible's emphasis upon the purposefulness of creation is especially intuitive. On the one hand, it encourages us, affirming what we all know and feel, that man does indeed have a purpose, and an exalted one at that. On the other hand, it is practical and illuminating, telling us much of what that purpose is: to know God, to co-labor with him, to raise a family, to serve as caretakers of his creation and to take dominion over the vast potentialities of the earth. On this crucial point the biblical cosmogony is far more intuitive and reasonable than either naturalism or pantheism.

The idea that God rested after his brief creative work certainly accords with intuition, so long as we remember that he is now represented as working in other ways, actively sustaining, animating, and guiding his every creature. What's more, the declaration of God's creation rest is well supported by the findings of modern science. Direct observation teaches us that no new matter, no new elements, no new stars, galaxies, planets, or life forms are coming into being. The basic shape of things is, as the Bible declares, fixed, though subject also to a mysterious principle of decay that threatens, in time, to destroy the integrity of all. Unlike cosmic evolutionism, the biblical model of creation (and fall) predicts what we actually see and is therefore more reasonable to believe.

The Age of the Universe, Life, and Man

We must, however, now linger over one feature of the biblical cosmogony that does indeed strike many people as ureasonable—the Bible's assertion that the universe was created in six literal days and that the world is therefore only some 6000 years old. How can this possibly be when, from many different scientific quarters, we repeatedly hear that the universe is 15-20 billion years old, the solar system 4.5 billion, life 1.5 billion, and *homo sapiens* around 1 million? What about all the geological evidences? What about radiometric dating? What about the billions of light-years required for starlight to reach earth from distant galaxies? With so much evidence stacked up against its teaching, surely the Bible must be in error on this fundamental point.

In beginning to wrestle with this important question, the seeker will want to keep in mind several key principles discussed earlier in our journey.

Above all, he must remember that in the matter of origins it is never wise to trust the claims of natural science over the declarations of a well-attested revelation. Natural science is blind to the beginning: it cannot observe the origin of the cosmos. But the creator is not blind, for he was there. If, then, we find in the world a clear revelation of the beginning—one that manifestly bears the (supernatural) *imprimatur* of the creator—it would be foolish and irrational indeed to spurn it in favor of the ever-changing claims of naturalistic cosmogony.[22]

The Bible, as we have seen, meets all these criteria. Its teaching about the beginning, unlike those of pantheistic religions, is crystal clear. Not only so, it is unanimously endorsed by all the biblical authors, including Jesus of Nazareth, the biblical Teacher *par excellence*. Also, it is supported by a vast array of supernatural signs—signs that undergird not only the authority of Jesus but the authority of the Bible as well. Here, then, is a book with all the earmarks of a trustworthy revelation. And if that book says the universe was created in six days, it would be foolish not to take the claim seriously.

Yet for all this, it is undeniably true that the seeker has a right to expect at least *some* scientific evidence favorable to a recent creation. How, indeed, could it be otherwise, if the God who *says* he created the world in six days really *did* create it in six days? Surely there would have to be at least some correspondence between the testimony of his book and the shape of his world. Surely, for the sake of all who must take his test, he would leave at least some evidences in the realm of nature to confirm his revelation of a recent creation.

The contention of biblical creationists is that God has done so, and done so very generously. Indeed, they argue that the *preponderance* of "cosmic chronometers"—natural phenomena indicating to us the age of things—supports recent creation. They claim, quite apart from the far more important biblical evidence, that scientific evidences *alone* will incline reasonable people toward a young universe rather than an old. All of this falls shockingly upon the ears of seekers schooled in the modern educational establishment. But as seekers they must give the creationists their due. They must take a little time to hear their arguments and consider their evidences. Only then can they decide for themselves which view of the age of the universe is more reasonable.

We must, then, briefly survey the creationist case.

A preliminary part of that case is a philosophical reminder. We must always remember, say the creationists, that men of science cannot help but bring certain presuppositions to their study of nature. One of them is the presupposition that present processes have continued in more or less the same way and at the same rate throughout history. Since the days of geologist Charles Lyell, this presupposition has been called *uniformitarianism*. Fundamentally, it is quite reasonable. Every day we observe uniformity in nature, whether in the motions of the heavenly bodies, the action of the tides, the behavior of animals, or the functioning of our own bodies. Without uniformity the world would be chaotic and science impossible. We are right to presuppose a significant degree of uniformity in our study of the past.

Nevertheless, we must be careful. For even on naturalistic grounds, we cannot assume that nature operates with *absolute* uniformity. Earth-shaking catastrophes may have interrupted the normal course of events in the past, powerfully influencing the present shape of things. Furthermore, from the moment we step into the biblical universe we must be even *more* careful. Yes, uniformity in nature harmonizes well with the biblical revelation of a rational God imposing a rational order on nature. But that same revelation also teaches us that nature has *not* always functioned in accordance with present-day processes. In the beginning, says the Bible, the world was suddenly and supernaturally created. Shortly afterward, it was suddenly and supernaturally cursed. Not long after that, it was suddenly and supernaturally engulfed in water and riven by a massive break-up of the earth's crust. Finally, there came a sudden and supernatural confusion of human speech, resulting in a dispersion of the family of man and the rise of distinct peoples and nations. Only after all this, says the Bible, did nature finally "settle down" into the more or less uniform set of processes we observe today.

Now if all this is true, we cannot possibly hope to understand cosmic history strictly in terms of present-day natural processes. Rather, we must see it not only through the lens of uniformity but also through the lens of these four divine interventions, each of which, if they really did occur, must surely have left its imprint upon the world as we now know it. When we do, say the creationists, many puzzling natural phenomena suddenly begin to make perfect sense. When we do, nature itself seems to confirm the biblical view, or model, of the beginning. And when we do, the world and the universe suddenly start to look very young.

But let us now observe these principles in action. We'll begin by looking briefly at a dozen of the best scientific evidences that creationists use to argue for a recent creation. Then we'll conclude by tackling the main evidences evolutionists use to argue for an ancient cosmos that is billions of years old. Interested readers may follow the footnotes to books that will take them deeper into this fascinating area of study.[23]

The Winding Up Dilemma

Astronomers have observed that stars in the spiral galaxies rotate around their galactic centers at different speeds, with the stars nearest to the center rotating much faster than those further out. This means that there is a built-in tendency for such galaxies "quickly" to disintegrate into a featureless cluster of stars. Our own Milky Way is such a galaxy. Because it is "winding itself up" so fast, yet still retains its spiral structure, scientists calculate that it cannot be more than a few hundred million years old, and may be less. It is certainly not 10 billion years old (byo), as evolutionists claim.[24]

Missing Supernova Remnants

A supernova is an exploding star. Astronomers observe that in galaxies like ours stars explode at a rate of about one every 25 years. Their remains should be visible for around a million years. If, then, as evolutionists argue, our galaxy is 10byo, we should be able to observe multitudes of supernovae in nearby portions of the Milky Way. In fact, we observe only about 200. This is about 7000 years worth of supernovae, and further evidence that our galaxy is, in fact, quite young.[25]

Our Shrinking Sun

Using direct visual measurements and several different indirect techniques, many scientists have concluded that the sun is shrinking, possibly by as much as five feet per hour. If this is so, temperatures on the earth only a few million years ago would have been so high as to destroy all life as we know it. A shrinking sun means that life could not possibly have begun to evolve 1.5 billion years ago, as evolutionists claim, and that the sun-earth system must be very young.

Here it is appropriate also to mention "the faint young sun paradox." Many scientists now believe that the sun is powered by nuclear fusion. On this view the rapidly moving nuclei of hydrogen atoms fuse to produce atoms of helium plus the release of huge amounts of energy. Because this process increases the density of matter in the sun, it leads to yet more fusion and the release of yet more energy. Here, then, is the paradox: contrary to expectation, the sun actually burns hotter and brighter as it gets older. But this implies that the sun, at its birth, must have burned cooler and fainter than it does today. And this, in turn, casts a long shadow of doubt over the antiquity of the earth-sun system, since the temperature on earth a billion years ago would have been far too low to sustain life.

Notably, these observations lead us to opposing conclusions. Is the sun shrinking, so that the earth was hotter in the past; or is it heating up, so that the earth was cooler in the past? Of the two, the former seems most likely, since the evidence for a shrinking sun is weighty, whereas many scientists remain uncertain about whether the sun is really powered by nuclear fusion. Whatever the truth may be, these observations show that the earth-sun system currently exists in a delicate—and continually changing—balance, a balance that can by no means accommodate billions of years.[26]

Solar Wind and the Poynting Robertson Effect

Scientists observe a cloud of small dust particles orbiting our sun (and other stars, as well). The smallest of these are being "blown away" by the sun's radiation. The larger ones, colliding with the same radiation, are first slowed and then "vacuumed" into the sun by its gravitational field. But unless the dust cloud is somehow being replenished (for which there is no evidence at all), a 4.5 byo sun should long ago have blown it away or vacuumed it up. Its continuing presence with us suggests that the solar system is actually quite young.[27]

Missing Meteorites

Meteorites, some of them quite large, accumulate on the earth's surface at a rate of about 60 tons per day. Yet there is little or no trace of meteorites in any but the topmost layers of the geological column. Since in the past meteorites probably fell more abundantly rather than less, their absence from the geological column suggests that the column did not form slowly over hundreds of millions of years, but quite suddenly, perhaps as a result of a global flood. Also, at present rates of accumulation, a 4 byo earth should have very large quantities of nickel-bearing meteoritic dust in its crust. The fact that it does not suggests that the earth is young.[28]

Decay of Earth's Magnetic Field

Direct measurements over the last 140 years have revealed that the strength of the earth's magnetic field is steadily decaying. Many believe this field is produced by an electrical current running through the molten core of the earth and that its decay corresponds to the cooling of the core. If so, the earth could not possibly be much older than ten to twenty thousand years since the heat generated by an older, larger current would certainly have destroyed its integrity. Also, evolutionists cannot explain how the earth, after 4.5 billion years, has not yet completely cooled but still has a molten core as well as a magnetic field that is presumably generated by it. Responding to all this, some evolutionists argue that the decay may only be apparent, an aspect of those periodic oscillations of the field that we know from geology to have occurred in the past. Creationists reply that the evidence for such oscillations indicates that they were quite rapid, possibly caused by events associated with the flood. They also point out that oscillations in the magnetic field cannot halt or explain its overall decay.[29]

Recession of the Moon

The moon is receding from the earth at a rate of about 1.5 inches per year. Even assuming that its recession began from earth and that the rate has been uniform, the moon could not possibly be more than 1.4 byo, a far cry from the 4.5 billion years assigned to the earth-moon system by evolutionists. Also, terrestrial life could not have appeared within the standard evolutionary time frame, since massive tidal activity caused by a moon so near the earth would have made this all but impossible.[30]

Atmospheric Helium

Most, if not all, of the helium in our atmosphere comes from the decay of radioactive elements in the earth. If the earth were 4.5 byo, there should be about 2000 times more helium in our atmosphere than we actually find. And since we know that very little helium escapes our atmosphere, a reasonable conclusion from all this is that the earth's atmosphere is young. Also, scientists have discovered pockets of helium located in deep, hot rocks that are thought to be about one billion years old. But since the helium is still there, it would appear that it has not had enough time to escape and that the rocks are really only several thousand years old. This finding not only implies a young earth but raises serious questions about the reliability of methods used to determine the age of rocks.[31]

Ocean Deposits

Each year a great deal more sodium enters the oceans than leaves it. Assuming that the oceans began with no salt in them, the present amount, measured against its present rate of accumulation, indicates a maximum possible age for the oceans of 62 million years. Similar measurements of oceanic copper, gold, lead, mercury, nickel, tin, and uranium give even younger ages. Yet evolutionists contend that the oceans have been around for some 3 billion years.

Much the same is true concerning the deposition of river sediments. These enter the oceans at a rate of more than 27 billion tons per year. Measurements of river sediments now on the ocean floor indicate a maximum possible age for the oceans of 30 million years. (If, however, the earth recently experienced a global flood, this figure would obviously be much too high.) Similarly, at their present rate of erosion, the continents would be completely leveled within about 25 million years. So again, it appears from both land and sea that the earth cannot be billions of years old.[32]

Intact Biological Tissue

Biological material decays very rapidly, and soft material more rapidly than hard (e.g., bone). This is why scientists have been shocked to find intact soft tissues in specimens thought to be millions of years old. The most recent example comes from the Hell Creek Formation in Montana, where evolutionist Mary Schweitzer discovered flexible blood vessels inside the fossilized thighbone of a Tyrannosaurus Rex (assumed to be 68-70 million years old). Not long before that, researchers found partially unfossilized dinosaur bones in the same area. Also, strands of the highly perishable DNA molecule have been recovered from a number of "ancient" specimens (e.g., from wet fossil magnolia leaves dated at 17-20 million years old). Finally, there is the case of Dr. S. Cano, who carefully extracted bacteria spores from the body of a stingless bee preserved in amber. To his amazement, he induced the "25-40 million year old" bacteria to grow! While scientists confess astonishment at such findings, they usually decline to embrace the most reasonable explanation: the intact biological material before their eyes is actually quite young.[33]

World Population Growth

According to evolutionists, the human race is about 1 million years old. If so, at a *very* modest growth rate of only .5% per year, world population today should be about 10^{2100}! Also, on this premise the geological column should be filled with the remains of multiplied trillions of human beings who walked the earth. It is not. On the other hand, assuming the same .5% growth rate, it would take only about 4000 years to produce today's population from a single couple. The evidence plainly favors a young human race.[34]

Human Culture and History

According to the usual evolutionary scenario, Stone Age men existed as hunters and gatherers for about 100,000 years, during which time they fashioned tools, erected simple dwellings, made stone monuments, painted on cave walls, and even made silk! Yet despite such displays of intelligence, it was not until about 10,000 years ago that they discovered agriculture and soon afterwards began to commit their stories and beliefs to writing. Thus evolutionists posit a great temporal gulf between the emergence of human culture and human history. Critics argue, however, that such a gulf seems highly improbable and that the more reasonable conclusion is that human culture and history began simultaneously, about 5000 years ago. This accords well with the biblical picture of early human history, and in particular with its story of a gradual re-emergence of human civilization after the devastating impact of the flood and the dispersion at Babel.[35]

These are only a few of the many processes cited by creationists as evidence for a young universe and a young earth.[36] Since all of them involve assumptions about the unobservable past, none can be said to give conclusive proof of a young earth. Nevertheless, it is highly significant that *approximately 90% of all cosmic chronometers give ages for the universe, earth, life, and man that are significantly younger than those proposed by evolutionists.* Indeed, the scientific evidence favors a universe so young as to rule out the possibility of cosmic evolution. The only other alternative is divine creation. Therefore, of the two options, divine creation appears to be the more reasonable.

But what of the much-publicized evidences for an ancient earth and cosmos? What of old-earth geology, radiometric dating and the problem of starlight and time? Is it really possible that these well-known cosmic chronometers are speaking to us erroneously about the age of things?

Yes, say the creationists, it is. To see why, let us very briefly examine each one.

Old Earth Geology

For the last 200 years, *uniformitarianism* has been the guiding principle of secular geology. According to this principle, the present is the key to the past. Just as geological change occurs slowly today, so it must have occurred slowly yesterday. The drift of the continents, the rise of mountains, the carving out of great canyons, the formation of the fossilized geological column—all, we are told, took place gradually over millions and even billions of years. For many, the presumed uniformity of geological processes conclusively demonstrates an ancient earth.

Today, however, many geologists have begun to acknowledge the inadequacy of the uniformitarian premise. This change is fueled by the dramatic discovery of several categories of new evidence, each of which suggests that the entire surface of the earth may well have been shaped by a catastrophic geological event, or series of events, that occurred in the recent past.

First, there are numerous evidences for the *rapid deposition of sedimentary rocks.* Chief among them are fossils. Their very existence in the geological column implies rapid deposition, since the original life-forms had to have been buried quickly and put under significant pressure in order to escape predation and/or decay. Interestingly, some fossils supply extraordinary snapshots of this very thing. Geologists have found, for example, fossils of highly articulated soft-tissue animals (e.g., jellyfish), an ichthyosaurus in the process of giving birth, one fish eating another fish, and tree-trunks shooting up through several layers of rock or coal (these are called *polystrate fossils*). Striking evidences like these confirm in the minds of many that *the entire fossilized geological column* may well have been laid down in a single massive hydraulic catastrophe such as the biblical flood.

Besides fossils, there are other evidences of rapid deposition. At the Grand Canyon, for example, geologists have discovered delicate imprints (e.g., animal tracks, ripple marks, rain-drop marks) on the tops of strata covered by other strata. Apart from rapid deposition, such imprints would surely have been worn or washed away. Again, on the tops of these strata there is usually no sign of embedded life (e.g., worms, roots, clams, etc.). But apart from rapid deposition, such life forms are very much to be expected. Or again, between layers of sedimentary rock geologists usually find no loose soil or signs of chemical erosion. Apart from rapid deposition, they certainly should. Uniformitarians are troubled by these phenomena. Creationists, on the other hand, are pleasantly surprised, seeing in them solid proof for a global flood that suddenly laid down the vast majority of earth's sedimentary layers.[37]

Second, we have evidences suggesting that *the earth's entire crust experienced a massive restructuring in the not-too-distant past.* Once again we think of the Grand Canyon, where, at the Tapeats, the horizontal layers of sandstone suddenly bend and rise toward the vertical within the space of 100 feet. Similarly, the entire Grand Canyon sequence of layers (allegedly deposited over the course of 300 million years) is bent at the Kaibab Upwarp. These layers show no sign of shearing and must therefore have been soft and pliable when, as a result of mighty forces from beneath, they suddenly rose to their present position and finally hardened.

The Sullivan River Mountains in British Columbia tell much the same story. Comprised of thick layers of unbroken sedimentary rock, these mountains look like interconnected hairpins. To view them is to understand immediately that in a relatively short space of time the soft layers of earth not only rose but were somehow compressed, accordion-like, into the mountain range we see today. Such amazing structures stand as a memorial to a brief but unimaginably powerful geological event.[38]

Biblically oriented geologists, observing phenomena like these, remember that the flood account speaks not only of the "opening of the windows of heaven" but also of the "breaking up of the fountains of the great deep" (Gen. 7:11). Some among them lay special emphasis upon the latter, theorizing that a sudden rupture in the basement rocks of the earth led to the explosive release of a vast reservoir of underground water. This globe-encircling rift led to a complete restructuring of the planet's surface in only a few short years. Indeed, they argue that this one event supplies the master key to historical geology, through which alone we may accurately understand such diverse phenomena as ocean ridges, trenches and canyons; continental drift, shelves, and slopes; mountain ranges, overthrusts, volcanoes, lava, metamorphic rock; and even comets, asteroids, and meteorites! Needless to say, such thinking occurs far outside the box of orthodox geology. Nevertheless, the growing evidence for a global catastrophe suggests that one day theories like these may well get the hearing they so richly deserve. If they do, the results for uniformitarianism could be catastrophic.[39]

Finally, we have a miscellany of evidences pointing to a recent global flood. We now know, for example, that vast animal graveyards have been found all over the world, revealing the sudden burial in mud of all kinds of beasts. Research in the Arctic and Antarctic has shown that both these regions once produced abundant vegetation in subtropical climates—until something happened. Fossilized marine crustaceans have been found on many of the highest mountains in the world. Entire skeletons of whales have been found in inland regions, and even on mountain tops! Pillow lava, which only forms under water, has been discovered on the peaks of Mt. Ararat. The earth's coal and oil deposits point to a flood since they could have formed only when large masses of organic material were suddenly submerged, sealed, and compressed in hot mud. The Grand Canyon points to a flood since only massive amounts of water rushing through soft layers of sediment could possibly have carved out so much material (the humble Colorado simply will not do). More than 230 legends from peoples all around the world point to a flood—legends which usually agree with the biblical record in all essentials: a global cataclysm, the destruction of mankind, a big boat, and a small family of survivors who went on to replenish the earth.[40] Finally, a global flood is implied by many sightings, reported over the last 150 years, of a huge, rectangular ship on the top of Mt. Ararat. Today many scoff, a few search, and the rest of us patiently wait to see if someone, somehow, will actually discover Noah's Ark.[41]

Summing up, we have seen that the uniformitarian premise of modern historical geology is not altogether unreasonable but that hard geological evidence actually favors a catastrophist interpretation of the earth's present geological structure. This evidence weighs heavily in favor of a recent creation, a young earth, and a devastating global flood.

Radiometric Dating

In various media, including text books, we are often told that radiometric dating has proven a given rock or fossil to be millions or even billions of years old. And yet, as we have already seen, the vast majority of dating methods suggest that the earth is quite young, possibly only thousands of years old. How can this conflict be explained? Is it possible that radiometric dating is not nearly so accurate as we have been led to believe?

To answer this question, we must understand how radiometric dating works and what assumptions are involved. We know that isotopes of certain elements such as uranium, potassium, and rubidium are unstable. That is, they tend to decay through the loss of atomic particles until they become another more stable element such as lead, argon, or strontium. We also know the (present-day) rates at which one such element decays into another. Many scientists, therefore, conclude that if we measure the ratio of a parent element to its daughter element in a given (igneous) rock, we can reliably calculate the age of the rock from the time it cooled—that is, from the time it was "created."

Now it is clear that in using this method scientists are making three crucial assumptions. Each is worthy of a closer look.

First, they must assume that there was no daughter element present at the beginning of the rock's history. But this is not necessarily the case. In fact, studies of modern lava flow have shown that parent and daughter elements do indeed appear together in "newborn" rocks. Certain knowledge of the original quantities of parent-daughter elements is humanly impossible. But without it, radiometric dating cannot be considered trustworthy.

Second, the scientists assume that the rate of decay has remained constant throughout the rock's history. But again, this uniformitarian premise is a risky business, especially if the world of nature has more than once been touched supernaturally by the hand of God. Suppose, for example, that rates of decay (along with the speed of light) were quite rapid during the creation week but began to slow when God rested. Or suppose that they were very high at the earliest stages of the flood and the "breaking up of the fountains of the deep" (when, perhaps, many isotopes were created amidst titanic geochemical transformations), after which they gradually began to slow. In other words, once we begin to interpret earth's history through the prism of the biblical paradigm, the uniformitarian assumptions underlying radiometric dating may legitimately be called into question.

Good science invites us to do so. Yes, measurements over the last 100 years indicate that rates of radioactive decay are now constant. However, recent researches have uncovered much evidence suggesting that this has not always been so. For example, geologists have time and again detected Carbon-14 in petrified wood, fossils, coal, etc. Since C-14, with its short half-life, should be undetectable after a maximum of about 50,000 years, one would think that the geologists would date the samples accordingly. But they do not. That is because the samples are found in rocks that other radiometric methods have "proved" to be millions or even hundreds of millions of years old. To retain the preferred older age, the geologists usually argue that the C-14 has somehow seeped into the old rocks. But surely it is at least as reasonable to argue that the "old" rocks may really be quite young because the rates of decay of their radioactive elements were faster in the past.

Again, rock samples taken from the same location—and, therefore, known to originate at the same time—have been assigned vastly different ages by different radiometric methods, (see below). Obviously these ages cannot all be right. Indeed, one reasonable conclusion from such anomalies is that *none* of them is right, since *all* the radioactive elements in the rocks decayed at an accelerated rate in the not-too-distant past. Why and when the acceleration occurred, the rocks do not say.[42]

Of special interest here is a recent study of zircons found in deep (Precambrian) granite. Zircons are crystals—in this case crystals containing lead produced from a uranium isotope. By measuring the amount of radiogenic lead in their samples, and by assuming the present rate of decay, researchers determined that these zircons should be about 1.5 billion years old. Yet other evidence indicated that this was not possible. One by-product of uranium-lead decay is helium, an element whose light, tiny atoms would quickly work their way up through porous crystal into the atmosphere. In other words, after 1.5 billion years, the radiogenic helium in these zircons should have been long gone. In fact, however, the researchers found large quantities of it. With the help of an independent analyst, they therefore established the exact rate at which the helium should have diffused from the zircons into the atmosphere. Then, using this figure, they determined from the volume of remaining helium the age of the rocks in which it was found: 5,680 (+/- 2,000) years.

These results are amazing. They strongly confirm what creationists have long suspected—that rates of radioactive decay were indeed much accelerated in the not-too-distant past. Moreover, the study of these deep basement rocks suggests that the earth itself may be only about 6000 years old—just as the Bible teaches.[43]

This brings us to the third and final assumption—that the rocks in which radioactive decay occurs constitute a closed system, immune to outside influences. Common sense tells us that this may be the most problematic assumption of all. All objects interact with their environments. Argon, for example, exists on earth as a gas that readily diffuses out of rock. Potassium and uranium dissolve in water. Heat, pressure, or chemicals may cause certain elements to "migrate" from one location to another. Such considerations again call to mind the flood. If a global flood did occur, it is inconceivable that it would not influence the parent-daughter ratios in all rocks.

We find, then, that there are a great many variables in the radiometric dating equation, any one of which could give spurious ages for rocks. Accordingly, it is not surprising that radiometric dating has time and again produced troubling anomalies—anomalies that are all too often glossed over by professionals in their public pronouncements.

Here are a few telling illustrations.

The potassium-argon method dated lava from the Mt. St. Helen's eruption at 350,000 years old. The same method applied to submarine basaltic rocks with a known age of 200 years gave ages ranging from 160 million to 3 billion years.

In one very extensive study, different methods gave vastly different ages for the same Grand Canyon rocks:

A. potassium/argon—Ten thousand to 117 million years

B. rubidium/strontium—1.27 to 1.39 billion years

C. lead/lead isochron—2.6 billion years

Rocks formed in two recent eruptions of the Mount Ngauruhoe volcano in New Zealand were assigned radiometric ages ranging from 270,000 to 3.5 million years.

Australian researchers found charred wood in a layer of Tertiary basalt. Carbon-14 dating put the age of the wood at 45,000 years, but potassium/argon dating put the basalt at 45 million.

Carbon-14, which should be undetectable after a maximum of 60,000 years, has been found in wood, coal, and fossil samples embedded throughout the entire geological column—even in rocks and layers thought to be hundreds of millions of years old.

C-14 has also been detected in diamonds. In one study, scientists discovered it in a South African diamond extracted from deep basement rock assumed to be more than 600 million years old. The diamond's carbon-dated age was put at 58,000 years. Note carefully that the dense lattice structure of this diamond virtually rules out the possibility of C-14 contamination from the outside. This means that the diamond must have formed from organic matter within the last 60,000 years or less. It also means that it must have formed under extraordinary—perhaps even catastrophic—conditions.[44]

Radiocarbon dating of a single family of mummified seals in southern Victoria gave dates ranging from 615 to 4600 years. Using the same method, one researcher determined that his living snails had died 27,000 years ago![45]

While the public is led to believe that radiometric dating is trustworthy, many scientists, aware of anomalies like these, privately admit otherwise. Dr. William Stansfield, a biology professor at California Polytecnic Institute, puts it this way:

> It is obvious that radiometric techniques may not be the absolute dating methods that they are claimed to be. Age estimates on a given geological stratum by different radiometric methods are often quite different (sometimes by hundreds of millions of years). There is no absolutely reliable long-term radiological 'clock.' The uncertainties inherent in radiometric dating are disturbing to geologists and evolutionists.[46]

It appears, then, that of all the cosmic chronometers we have looked at, radiometric dating is among the least reliable. It requires a knowledge of past conditions that scientists do not have and involves a uniformitarian interpretation of earth history that, even among secular geologists, is highly suspect. Those who seek to know the true age of the earth will not, therefore, be unduly influenced by this method.

Starlight and Time

A third cosmic chronometer suggesting an old universe is the phenomenon of starlight. The argument here is deceptively simple: if, as many astronomers now assume, distant stars and galaxies are indeed billions of light-years from earth, then the universe must be billions of years old. If it were not, how could the light from those bodies have had sufficient time to reach us?

As I say, this argument is deceptively simple since it involves a number of crucial assumptions, none of which is demonstrably true. For example, the argument assumes that the starlight we are now seeing did in fact originate in a star or galaxy. It also assumes that the object really is billions of light years away. It assumes that the light coming from the object has traveled at the same speed throughout its journey to the earth. And it assumes that the space through which the light travels does not in any way affect the way we perceive it here on earth. Scientists can be found to challenge all these assumptions. More importantly, the challenges themselves underscore a crucial fact that we have met again and again in our study: the history of the universe (and therefore the truth about these assumptions) cannot be known with absolute certainty by the empirical methods of natural science. A divine revelation is required, or we are shut up to ignorance and guesswork about the pre-human past.

But let us look at each of these assumptions a little more closely.

Consider first the assumption that distant stars and galaxies really are billions of light-years away. This is far from certain. The trigonometric method by which we may fairly accurately measure astronomical distances is useful only to a limit of between two and three hundred light-years from the earth. Beyond this we are reduced to educated guesses based on the size, intensity, and red shift of the bodies in question. Among these, the last is by far the most important since it is widely believed that the red shift of a star or galaxy is directly proportional to its distance from an observer. We have seen, however, that there is growing scientific uncertainty about the true meaning of red shifts. As a result, there is growing uncertainty about the true size and age of the universe. Yes, many astronomers think the most distant galaxies are about 15 billion light-years away. Others, however, propose that they may be as close as several hundred! Only one thing is sure: no one knows for sure.

Second, there is the assumption that the speed of light (c) has remained constant throughout cosmic history. This too is far from certain. Dr. Walt Brown reports that over the last 300 years some 160 measurements of c indicate that it is decaying steadily. One Russian cosmologist, V. S. Troitskii, infers from the evidence that at the beginning of the universe c may have been 10 billion times faster than it is today! Many modern cosmologists are intrigued with the thesis of c-decay. If verified, they believe it would go far toward explaining such puzzling phenomena as galactic red shifts, the cosmic background radiation, anomalous ages given by radiometric dating, and other cosmological problems associated with the Big Bang.[47]

But even if it turned out that c is not gradually decaying, other interesting possibilities remain for those open to the biblical testimony. Perhaps, for example, c was very high throughout the fourth, fifth, and sixth days of the creation week but then slowed to its present (constant) value on the seventh day, when God rested. Or perhaps, as some have suggested, c was influenced by the curse that fell upon all nature at the time of Adam's sin. On this view, c would gradually decay and then, at some point in the past, settle down to a norm, rather like the life span of humans over the first 2000 years of cosmic history (Genesis 11). The Bible does not teach any of these views, but they all are consistent with its cosmogony and draw upon it for inspiration. Naturalists are not likely to consider them. Anyone open to biblical revelation will consider them with care.

Next, there is the assumption that light from the stars travels through space in a straight line. As natural as it is to believe this, some astronomers argue that it may not be so; that space is actually hyperbolic, or "curved;" that it bends light rays so as to make objects appear farther away than they actually are. In other words, if space is hyperbolic, the universe may in fact be a sphere several *thousand* light-years in diameter while appearing to be a sphere of several *billion*. Though this model of the universe is fairly counterintuitive, there does appear to be some observational evidence in its favor.[48]

Finally, there is the assumption that the light we see actually emanated from the objects we see—objects presumed to be billions of light-years away. Now in the universe of the naturalist, this is a perfectly reasonable assumption. But once we allow the Bible to speak to the issue, another possibility, just as reasonable, presents itself—that God, on the fourth day, simply "switched on the stars," creating not only the heavenly bodies themselves, but streams of light connecting them with the earth (and one another). In other words, it is possible that he instantaneously created a mature, fully functioning, earth-related astronomical heaven—a heaven that had, as a necessary by-product, an *appearance* of old age.

Though the Bible does not explicitly teach this view, it is, in the minds of many creationists, the most biblical of all explanations for the apparent old age of the universe. Two main arguments may be cited here.

First, we know from Genesis 1:3-5 that on the first day God created an ambient light to bathe the earth in daytime. Could it be, then, that on the fourth day this light was somehow distilled not only into the stars but into streams of light connecting them with the earth? If so, starlight would effectively have been created *en route* to, and present at, the earth—with no need of time for it to travel from a physical source.

Second, we learn in Genesis that all earthly things—the land, sea, sky, plants, fish, birds, and animals—were, in essence, created instantaneously. This entails that they were also created mature and fully functioning, ready to serve the soon-coming ones for whom they were made: the family of man. And this entails that all things were created with an appearance of age, though in fact they were fresh from their Maker's hand. Would not the same, then, be true for the heavens and their contents? Would they not also have been created "up and running," linked by light to the earth, ready to serve man as luminaries, signs, and markers of time? And would they not (to a modern scientist, in any case) have given an appearance of *very* old age, when in fact they and their light were altogether newborn?

Here, say the creationists, is the great value of having God's revelation: it supplies a true model, or paradigm, through which we may accurately understand what we see around us, including how and when it came to be. Yes, to a biblical illiterate, starlight (and other natural phenomena) might indeed give the universe an appearance of old age. But God has graciously supplied a revelation so as to prevent or correct that (false) impression. Men are, of course, free to disregard this revelation in favor of their own (naturalistic) theories. But in that case they have none but themselves to blame when the appearance of things leads them to wrong conclusions about their origin and age—and much else besides. To judge from their difficulties in pinning down the age of the universe, such would appear to be the case among most modern cosmologists.[49]

In closing, it must be stressed yet again that the Bible does not reveal precisely how God created the stars or the light connecting them to the earth. Perhaps the universe is really very small. Perhaps, in the past, light traveled at near-infinite speed. Perhaps space is curved. Perhaps cords of light joined the newborn earth to the newborn stars. *And perhaps the truth involves two or more of these possibilities.* Such uncertainty may seem disappointing, but we must remember: the value of considering these suggestions does not consist in trying to tease out of nature what God has not been pleased to reveal in Scripture. Rather, the value consists in helping seekers understand that it is actually quite reasonable to believe the biblical account of the beginning—and to feel sure that there is indeed a good explanation for the as-yet unresolved mystery of starlight and time.[50, 51]

In our evaluation so far we have seen that the biblical cosmology is readily understood, highly intuitive, logical, and well-supported by scientific and historical evidences—far more so than its naturalistic or pantheistic counterparts. Perhaps, then, the 45% of Americans who still embrace it are not so unreasonable as their opponents would have us believe. If so, careful seekers must not let the dogmatism, ridicule, or invective of the scientific establishment deter them from the hard work of carefully investigating all sides of the great debate.

Is the Teacher's cosmology "right"? That is, does it supply a convincing rationale for our deeply felt convictions about an objective moral order? Most would say that on this point the biblical teaching is especially satisfying.

Above all, the biblical cosmology revolves around a personal God—a God who is altogether good and whose original creation was altogether good. Such affirmations deeply resonate with our ethical intuitions, supplying a solid theological rationale for our most fundamental impressions of the universe: that it is good, that it should be better, and that one day it will be better. Accordingly, for many this revelation becomes the ground of an idealism and optimism that enables them to work hard for a better world even as they wait for

God to bring in a perfect one. If Israel had not given the world such a cosmology, surely our own intuitions about the universe would drive us to invent one very much like it.

We observe also that the biblical cosmogony supplies a clear theological base for many of mankind's most cherished ethical norms. For example, it grounds our innate conviction about the sanctity of human life, teaching that man really is a being set apart; that he alone, among all God's creatures, is fashioned in God's own image and likeness; and that he, therefore, does indeed enjoy a (God-given) right to life and liberty, the right use of which is man's only hope for securing the happiness he so avidly pursues.

It confirms our intuition that man is a purposeful being with God-ordained work to do and goals to reach.

It grounds our abiding sense that man, in his God-ordained development of the earth, must carefully steward nature and lovingly watch over its animal life.

It sharpens our sense for healthy sexual relations, explicitly teaching that it is not good for men and women to live alone; that heterosexual marriage is divinely ordained for companionship, procreation, and teamwork; that in a marriage each partner has distinctive roles and responsibilities; and that husband and wife must be faithful to one another all their days.

In sum, we find over and again that the biblical revelation of how things *came* to be has much to teach about how they *ought* to be.

Is this cosmology hopeful? Eminently—chiefly because it posits a good creator whose original purpose was to be a Father to his free creatures. He is, quite clearly, a God who joys in the joy of his children. Seeing, then, that he also is an almighty creator, we take hope, knowing that he who purposed our joy in the beginning is doubtless well able to bring it to pass in the end. This is why suffering humanity will always greet the biblical cosmogony with keen interest—and why it cannot pass away. The biblical creation story, unlike all others, is actually a whispered promise of redemption. A creator such as this cannot fail. He will bring his children back; he will meet them in the garden; he will walk with them again.[53]

NOTES

1. For a table indicating the extent to which Jesus and his apostles referred to the beginning, see Appendix 3.

2. Some interpreters find a "cosmological trinity" in Gen. 1:1, arguing that in the beginning God created *time* (at the beginning), *space* (the heavens) and *energy/matter* (the earth). But if, as I suggested earlier, time is essentially a supernatural reality—and perhaps an attribute of God himself—it cannot be part of his creation. It seems best, then, to take the text at face value: in the (six-day) beginning of the universe, God created space and all that was meant to fill it.

In passing, we may also note here that the weight of biblical evidence favors the idea that the universe is finite. Everything else that God created was finite: why should space be the exception? In creation, God set fixed boundaries for the seas: why would he not set them for the heavens as well (Psalm 104:7-9, Jeremiah 5:22)? If he measured heaven with a span, must it not be finite (Isaiah 40:12)? If God knows the number (and the names) of the stars, that number is finite (Psalm 147:4, Isaiah 40:26). But if the number of the stars is finite, ought not their abode to be finite as well? Most importantly, God seems jealously to reserve the attribute of infinity to himself, explicitly declaring that he has no equal and that there is none like him (Isaiah 40:25, 46:5). If, then, God alone is infinite, how can the universe (by being infinitely extended) be his equal?

3. Some have argued that chapters 1 and 2 of Genesis represent separate and contradictory Hebrew creation stories, brought together at a late date by the unknown author/editor of Genesis. There is, however, no historical evidence to support this view, while the entire weight of Jewish tradition, as well as the testimony of Jesus and his apostles, dispute it by affirming the Mosaic authorship of Genesis (see Mark 12:26, Luke 24:44, Acts 28:23). A careful reading of these chapters will reveal that the accounts are not contradictory but complementary. Genesis 1 is an overview of creation as a whole; Genesis 2 further elaborates, focusing on the creation of man and woman, the institution of marriage, and the probation of man in the Garden of

Eden. (See Douglas Kelly, *Creation and Change* (Mentor Books, 1997), pp. 52-54, and *Revised Answers Book*, pp. 47-48.).

4. In his second letter to the Corinthians, Paul writes, "I know a man in Christ who fourteen years ago—whether in the body I do not know, or out of the body I do not know, God knows—such a man was caught up to the third Heaven... caught up into Paradise," (2 Corinthians 12:2-4). Here, Paul alludes to three heavens. According to many interpreters, the first is the atmospheric heaven, the second the stellar heaven, and the third the spiritual Heaven—the place of God's continuous self-revelation to the angels. As this passage suggests, the exact nature and whereabouts of the third Heaven (or Paradise) is not so easy to determine. We shall consider some possible answers further on.

5. There is much difference of opinion about the nature and whereabouts of the "waters which were above the expanse" (Genesis 1:6-9). Ancient interpreters viewed them as spiritual—the "sea of glass" located in Heaven just beyond the sphere of the fixed stars (Rev. 15:2). Others have suggested that they are physical, serving as the outer boundary of a finite cosmos that may, in fact, be much smaller than we imagine. Others say the waters were simply the first clouds.

A popular modern view is that these waters were something historically unique: a canopy of water vapor that surrounded the earth from its creation up to the time of the flood. Henry Morris points out that this hypothesis goes far toward explaining a number of important biblical and natural phenomena. These include the (apparent) lack of rain before the flood (Gen. 2:5, 7:4); the diurnal mist (or springs) that watered the antediluvian lands (Gen. 2:6); the waters that fell for forty days and nights when, at the time of the flood, "the windows of heaven were opened" (Genesis 7:11, 8:2); the late appearance of the first rainbow (Gen. 9:13f); the tropical climate of the ancient earth (including the Arctic and Antarctic regions), presumed to be the result of a greenhouse effect induced by the vapor canopy; the great longevity of antediluvian man (Gen. 5); and the greater size of most animals prior to the flood. (See Morris, "Let the Word of God Be True," *Acts and Facts*, January, 2003.)

It should be noted, however, that capable creationist critics have found this view both scientifically and biblically wanting. One of them, Dr. Walt Brown, therefore proposes that the expanse (or firmament) was actually a "plate"—a thick crustal sheath of earth. The waters beneath the plate (and upon which it floated in the beginning) were "the fountains of the great deep," (Gen. 7:11). The waters above it were the primordial seas, out of which, in places, the crust arose to form dry land. While Brown capably defends this view, it seems to break against Gen. 1:8a, where it is written, "And God called the expanse heaven (or the sky)." (An alternative translation, "Also, God called heaven the expanse," is grammatically forced.) Note also that this view shatters the apparent structure of Genesis 1, according to which the environment *formed* on day two should be *filled* with creatures made on day five. It will hardly do, however, to have birds flying "above the earth across the face of the plate" (Gen. 1:20, NKJV). (See ITB, *op. cit.*, pp. 260-268.)

6. In Ecclesiastes 3:21 (NIV), we find Solomon asking, "Who knows if the spirit of a man rises upwards and if the spirit of the animal goes down into the earth?" Though he cannot tell where it goes, Solomon definitely knows the animal has a spirit.

7. Though God is the giver and sustainer of life, the Bible does not teach *panentheism*—the idea that he alone is the indwelling principle animating all living matter. Rather (in the case of men and animals), it is the soul that animates matter—and, in a manner unexplained by the Bible, God who animates the soul, while remaining ontologically separate from any evil in it (Gen. 2:7, James 2:26).

8. It is noteworthy that the apostle Paul often appealed to the beginning when he supplied Christians with ethical guidelines for relations between the sexes (1 Corinthians 11:1f, Ephesians 5:22f, 1 Timothy 2:8f).

9. Bible-believing scientists of the past include such notables as Isaac Newton (physics), Johann Kepler (astronomy), Robert Boyle (chemistry), Lord Kelvin (thermodynamics), Louis Pasteur (bacteri-

202 · In Search of the Beginning

ology), Matthew Maury (oceanography), Michael Faraday (electro-magnetics), Clerk Maxwell (electrodynamics), John Ray (biology), and Carlous Linnaeus (taxonomy). See Morris, *The Biblical Basis for Modern Science*, p. 30; and also pp. 463-5, where he lists more than 60 outstanding creationist scientists.

10. Numerous biblical passages depict the world as a theatre in which man is tested and observed by powers beyond his ken (see 2 Chronicles 6:9, Job 1-2, Psalm 14:2, Matthew 18:10, 1 Corinthians 4:9).

11. Dr. Pattle Pun, a progressive creationist Christian, speaks for many when he writes, "It is apparent that the most straightforward understanding of the Genesis record, without regard to hermeneutical considerations suggested by science, is that God created heaven and earth in six solar days, that man was created in the sixth day, and that death and chaos entered the world after the Fall of Adam and Eve."

Nevertheless, says Pun, the scientific evidence amassed to support the theory of natural selection and the antiquity of the earth is so impressive as to trump the straightforward biblical teaching. (P. Pun, "A Theory of Progressive Creationism, *Journal of the American Scientific Affiliation*, (March, 1987):14.)

12. For a discussion of interpretations of Gen. 1-11 that seek to preserve a high view of the Bible while accommodating modern scientific opinion about the antiquity of the universe, life and man, see Henry and John Morris, *The Modern Creation Trilogy: Scripture and Creation* (vol. 1), (Master Books, 1997), pp. 35-64.

13. Conservative interpreters understand that as a matter of historical fact, if not logical necessity, the doctrine of an ancient universe goes along with cosmic evolution; that evolution goes along with the denial of the first man, Adam; that the denial of Adam goes along with the denial of an original sin (i.e., the sin which let evil, suffering and death into the cosmos); and that the denial of an original sin goes along with the denial of a need for a Savior—a last Adam who will undo all that the first Adam did. In other words, conservatives believe that in the battle for the beginning, the Gospel itself is at stake.

It is interesting to note that the opponents of Christianity some-times understand this better than its friends. Secular humanist Richard Bozarth is a case in point. He writes:

> Christianity is—must be!—totally committed to the spe-cial creation as described in Genesis, and Christianity must fight with its full might, fair or foul, against the theory of evolution… It becomes clear now that the whole justification of Jesus' life and death is predicated on the existence of Adam and the forbidden fruit he and Eve ate. Without the original sin, who needs to be redeemed? Without Adam's fall into a life of constant sin terminated by death, what purpose is there to Christianity? None! What all this means is that Christianity cannot lose the Genesis account of creation like it could lose the doctrine of geo-centrism, and get along. The battle must be waged, for Christianity is fighting for its very life. G. R. Bozarth, *The American Atheist,* Sept. 1978, pp. 19, 30.

14. A possible exception to this general rule is the ongoing cre-ation of the spirits of living beings at the time of their conception (See Psalms 104:27-30, 139:13-16).

15. For a rich meditation on the meaning of the Sabbath day, see Douglas Kelly, *op. cit.,* pp. 237-252.

16. Christ does not appear in Heaven by way of vision only. The Bible teaches that he (like Enoch and Elijah before Him) was carried into Heaven bodily, where he now appears in bodily form (Acts 1:9-11; Genesis 5:21-24, 2 Kings 2).

17. In this section I have defended the idea that Heaven is a mov-able visionary world existing within the physical cosmos. There are, however, other views. James Jordon, for example, argues that Heaven is a spiritual realm existing just beyond the edge of our finite universe, on the far side of the firmament (Genesis 1:6-8). Luco Van Den Brom and Hugh Ross opine that Heaven is an unimaginable spiritual di-mension, or set of dimensions, within which our three-dimensional cosmos is embedded. Ross's view, however, owes more to modern string theory than it does to the Bible. For further discussion see Byl, *op. cit,* pp. 161, 169-170, 206-209.

18. Jesus' teaching here would, of course, be one of the revelations that the Father desired him to bring into the world (John 7:16).

19. In the Revelation, Christ also refers to himself as "the Beginning of the creation of God" (Rev. 3:14). Some have interpreted this to mean that he is identifying himself as the first (angelic) creature that God made. But because so many other passages clearly designate Christ as the creator of *all* things, this view is impossible. The word here translated as "Beginning" (Greek, *arche*) can also mean *origin* or *source*. Translating it thus, the passage has the glorified Christ identifying himself as the creator of the universe.

20. The creation was a thoroughly trinitarian event, since, upon the Son's command, the Spirit "did the work" (Genesis 1:2).

21. For further such quotes, see Demise, *op. cit.*, p.127.

22. Even evolutionists recognize that "scientific" determinations of the age of the universe, life, and man are continually changing and, therefore, uncertain. A. Engel, an evolutionary geologist, is refreshingly candid on this point: "The fact that the calculated age of the earth has increased by a factor of roughly 100 between the year 1900 and today—as the accepted "age" of the earth has increased from about 50 million years in 1900 to at least 4.6 eons today—certainly suggests that we clothe our current conclusions regarding time and the earth with humility" (See MCT, *op. cit.*, p. 286.)

23. The evidences cited were drawn from four publications: 1) Russ Humphreys, "Evidence for a Young World," (an *Impact* article published by ICR, June, 2005); 2) Demise, pp. 51-76; 3) ITB, pp. 5-42; 4) MCT, pp. 285-336. See also the important new book, **Thousands, Not Millions**, By Don De Young (Master Books, 2005).

24. Evidence for a Young World, *op. cit.*, p. 1.

25. Demise, *op. cit.*, p. 65; ITB, pp. 27, 72.

26. ITB, *op. cit.*, pp. 34 and 82 for documentation concerning our shrinking sun. See also J. Sarfati, *Refuting Compromise*, pp. 169-171 concerning the faint young sun paradox.

27. *ITB*, p. 34.

28. Demise, *op. cit.*, p. 66.

29. Jonathan Sarfati, "The Earth's Magnetic Field: Evidence that the Earth is Young," *Creation Magazine*, March, 1998, pp. 15-17.

30. ITB, *op. cit.*, p. 33.

31. Evidence, *op. cit.*, p. 1.

32. Ibid., p. 1.

33. Also, Frank Sherwin, "The Devastating Issue of Dinosaur Tissue," (an Acts and Facts article published by ICR, June, 2005.) See Revised Answers Book, *op. cit.*, pp. 246-247.

34. Demise, *op. cit.*, p. 65.

35. Evidence, *op. cit.*, p. 1.

36. For lists containing about 80 more cosmic chronometers, see Demise, *op. cit.,* pp. 68-69, and the appendix entitled "Global Processes Indicating Recent Creation" in Henry Morris, *The Biblical Basis of Modern Science* (Baker Books, 1984), p. 477-479.

37. The material for this section is drawn from John Morris, *The Young Earth*, (Master Books, 1994), pp. 93-118.

38. See ITB, *op. cit.*, p. 94.

39. Much creative work along these lines has been done by Dr. Walt Brown. His *magnum opus, In the Beginning: Compelling Evidence for Creation and the Flood, (op. cit).*, is, among other things, a marvel of geological theorizing based upon the Bible and science. In this layman's opinion, no open-minded geologist can afford to neglect it.

40. ITB, *op. cit.*, pp. 41 and 83.

41. See Demise, *op. cit.*, pp. 53-58; ITB, pp. 37-41.

42. For a proposed explanation of the acceleration of rates of radioisotope decay, see Sarfati, *Refuting Compromise*, pp. 382-383.

43. See the article by Carl Wieland, "Radiometric Dating Breakthrough," in *Creation Magazine*, March-May, 2004, pp. 42-44.

44. It is important to remember that 60,000 years is an upper limit for the age of the diamond. If (as the diffusion of helium from

zircons suggests) the C-14 decayed at a higher rate in the recent past, its true age would be less.

45. These examples of anomalous radiometric dates, and others like them, are discussed and annotated in Demise, *op. cit.*, pp. 60-63, and The Revised Answers Book, *op. cit.*, pp. 75-94.

46. Demise, *op. cit.*, p. 60-63.

47. For further discussion on this fascinating and important topic, carefully read ITB, *op. cit.*, pp. 232-237. See also Douglas Kelly, *Creation and Change*, (Mentor, 1997), pp. 137-158.

48. See Byl, *op. cit.*, pp. 193-4.

49. For an extended discussion of mature creation, with responses to common criticisms, see Byl, *op. cit.*, pp. 194-201.

50. In writing this section I have relied heavily upon an excellent short article by Richard Niessen, "Starlight and the Age of the Universe." It appeared in the periodical, *Impact* (Institute for Creation Research, July, 93).

51. Evolutionists cite other natural phenomena in support of millions of years: varves, evaporites, fossil graveyards, fossil forests, bio-deposits, chalk deposits, coral reefs, rates of cooling in granite, rates of erosion, etc. By and large, these phenomena are easily explained with reference to a global flood, for which there is abundant evidence. For an extended discussion of each one, see *Refuting Compromise*, pp. 367-388.

52. For further reading on biblical creationism, see Henry Morris, *The Biblical Basis For Modern Science* (Baker Books, 1999); Duane Gish, *Evolution: The Fossils Still Say No!* (ICR Books, 1995); Duane Gish, *Creation Scientists Answer Their Critics* (ICR Books, 1993).

53. The hope glimpsed in the biblical creation story is made explicit in many other passages of Scripture. Fittingly enough, one of the most beautiful is found near the end of the book that brings the Bible to a close.

> And I (John) saw a new heaven and a new earth, for the first heaven and the first earth had passed away. Also, there was

no more sea. Then I saw the holy city, New Jerusalem, coming down out of heaven from God, prepared as a bride adorned for her husband. And I heard a loud voice from heaven saying, "Behold, the tabernacle of God is with men, and He will dwell with them, and they shall be His people, and God Himself will be with them and be their God. And God will wipe away every tear from their eyes; there shall be no more death, nor sorrow, nor crying; and there shall be no more pain, for the former things have passed away." Then He who sat on the throne said, "Behold, I make all things new." And He said, "Write, for these words are trustworthy and true."

— Revelation 21:1-5

CHAPTER 5

The Way to the Beginning

In our search for the beginning, we have come upon much information—information from the realms of history, philosophy, science, and religion. Such information is necessary. The bridge to the beginning is built out of information—good information, even divine information.

But it is clear that information alone is not enough. If it were, our information-rich age would have produced a consensus about the beginning. Since it has not, the discovery of cosmological truth must depend on something more. Drawing upon some painful personal experience, I want to spend a few moments at the end of our journey discussing what I believe the "something more" to be.

In the early months of 1970, as I continued my search for spiritual reality, I met weekly with Father Gabriel Barry to dialogue about Christianity and Roman Catholicism. His preferred method of catechesis was simply to let me ask questions. I would raise an issue, he would answer as best he could and, if necessary, loan me books that explored in depth the topic at hand. By this method we had soon touched upon all the questions of life. And by it I soon realized that biblical religion was different from eastern religion. Fundamentally different. A crisis was brewing. Though I greatly desired to bring my spiritual quest to a satisfying conclusion, I now began to see that becoming a Roman Catholic Christian would mean abandoning my pantheism and—at the deepest level of our relationship—my pantheist friends. Furthermore, it would also mean embracing certain biblical doctrines that I found incredible, frightening, and even repellent. For the first time my search for spiritual reality was becoming difficult and costly. Though I did not understand it then, my love of the truth was being put to the test.

During this challenging season, Father Barry and I discussed the question of origins. I wanted to know the official Roman Catholic position on evolution. In response, he gave me some books by Catholic theologians that, in essence, endorsed theistic evolution. Because of my own childhood indoctrination into evolutionism, I found this answer to be reasonable. Who could seriously question the fact of evolution? If God exists, he *must* have used evolution to "create" the cosmos.

More importantly, in time I also began to see in this answer a convenient solution to the apparent conflict between the Bible and eastern religion. For if the Bible did not speak clearly about the beginning (as Genesis certainly did not if cosmic evolution were true), then perhaps it also did not speak clearly on other matters—e.g., the nature of God, man, sin, Christ, salvation, and the afterlife. Perhaps the teachings of Jesus of Nazareth had a deeper, mystical meaning. Perhaps, as some asserted, he really did travel to India in his youth. Perhaps he really was a Hindu adept, a *boddhisattva*, an enlightened Master—the greatest of all time, no doubt, but one among many, nonetheless. In short, if the Bible spoke metaphorically about creation, perhaps it spoke metaphorically—and pantheistically—about all the rest.

In *The Test* I relate how things finally turned out for me. Here, however, I wish to make a confession that bears heavily on my theme in this book: the love of the truth and the search for the beginning.

I have often asked myself: If Father Barry had responded to my question about evolution differently; if he had defended the plain sense of Genesis; if he had supplied me with thoughtful books written by creationist authors; if, indeed, he had given me the book that I have now given you—would it have made any difference? Would I have turned from my pantheism to one or another form of orthodox (i.e., theistic and trinitarian) Christianity?

Such a question may be impossible to answer, but I will try anyway. With all the benefit of hindsight, and as best I understand my spiritual condition at that time, I do not think it would have made any difference. Yes, the unknown god was definitely drawing me to an investigation of Christ and the Bible. And I was genuinely interested in discovering spiritual truth. But I was also reluctant. Pantheism was still new and exciting to me. I had invested much time and energy in it. My blossoming spiritual identity was wrapped around it. Its promise of enlightenment gave focus to my existence. It lay at the center of my most significant relationships. How could I summarily abandon it now?

Meanwhile, Christianity, for all its attractions (the chief of which was Jesus himself), increasingly seemed foreign and threatening. It said certain things I did not want to believe. It required certain decisions I did not want to make. And so, because I did not want to believe or submit, I easily found reasons to do neither. Yes, Father Barry's answer to my question about origins was defective. But I do not think more or different information would have made any difference. What I really needed was more honesty. But because, at that time, honesty was in short supply, I eventually abandoned my catechism and re-immersed myself in the world of Zen.

Today, more than thirty years later, writing the last chapter of a book on cosmology, I therefore put myself in Father Barry's shoes. I imagine a seeker sitting before me, a young man who has read my book, a man such as myself those many years ago. He has just told me that he enjoyed the book and found it informative. Still, he is not persuaded. He needs more time to be alone and think. He simply cannot believe that the biblical beginning is true—or any of the other teachings of the orthodox Christian faith. He is grateful for our time together, but must temporarily break off the relationship. He will try to get in touch again later, when he has finally gotten things sorted out for himself.

I look at him, my heart sinking, knowing that he has no intention whatsoever of getting in touch later; knowing all too well the pain that awaits him, yet understanding also that he will have to go through that pain in order to find the truth.

What do I say to him? What parting words would best serve to illumine his path and ease his way?

Here they are. Though focused on the question of the beginning, I believe they apply equally well to all the other questions of life.

Dear friend, thank you so much for taking the time to study with me. I know of little that is more rewarding than to think deeply with a young seeker about one of the great questions of life. Being a part of your search for truth has been a high privilege and a deep joy. It was a gift to me. Thank you for it.

And now, as you head off on the next leg of your journey, please allow me to leave you with a few final thoughts. You have heard them before, but I believe they are important enough to merit a brief repetition.

Above all else, please continue to consider the proposition that your life is a test—a test of your love of the truth, set before you by a wise and loving unknown god. As I've told you elsewhere, there are many good reasons to believe this is so. And from the moment you do believe it is so, you will find that your life is suddenly framed, focused, and charged with new meaning. If you have been won to the test perspective, listen to it carefully: it will tell you everything you need to do. What's more, it will give you the hope and confidence to do it.

It will tell you, for example, that you must have *faith*. For if the unknown god has deposited within you a hunger to behold the beginning, obviously he means to satisfy it. You must, therefore, believe that somewhere out in the world he has graciously unveiled the truth about the origin of the universe, life, and man. Like wheat amidst chaff, or gold beside pyrite, it is partly hidden. But you must believe that somehow he will enable you—and every sincere seeker—to recognize it when you find it. You *can* find the truth, and you *can* be sure that it *is* the truth. But to do so you must believe that truth and assurance exist, and that they are waiting to be found by the one who sincerely seeks.

The test perspective will also tell you to be *diligent*. A test would not be a test without difficulty; and where there is difficulty, there must be hard work to overcome it. In the pursuit of truth, this means that you cannot run away from study, controversy, or confusion. Nor can you allow laziness to persuade you to defer to the judgment of experts. The experts may be right. On the other hand, in a world designed to test the lovers of truth, they may well be wrong, as history abundantly demonstrates. Thus the only way you can be certain about the beginning is to work hard to find it for yourself.

The test perspective also teaches you the importance of *self-confidence*. By this I do not mean the kind of confidence that says, "I have no need of anyone, human or divine, to help me. I can do it myself." To the contrary, I mean the kind of confidence that says, "An unknown god has created me. He has given me spiritual common sense, reason, ethical intuition, and an inclination to hope for the best. He obviously desires me to use these faculties in my search for truth about the beginning. And I am confident that if I do, neither they nor he will let me down—no matter how humble my intellectual gifts." This is godly self-confidence, healthy self-confidence. Because it is rooted in faith in a good and reasonable god, it means that you can boldly spurn all non-sense, irrationality, mystical double-talk and moral compromise. Yes, at times you will be appalled by your own confusion and foolish mistakes. But you will still remain confident, knowing that he who made your faculties is well able to compensate for their defects, so long as your intention is pure. You can trust that when truth appears, you will indeed recognize it for what it is.

Finally, the test perspective teaches you the necessity of *courage*. In the face of two great challenges, it will definitely be needed.

First, your pursuit of truth may require you to stand alone—and therefore courageously—before men. Or to state the case more precisely, it may require you to stand virtually alone with a despised minority. In other words, social marginalization may well belong to the essence of the test of life. This makes sense. If a test is to count for something, it must cost us something valuable. And what (we often think) could be more valuable than one's standing in the world—one's position and reputation in the great pecking order of human society? Ought we not, then, to imagine the divine Tester looking down upon a seeker, asking himself, "I wonder if he loves the truth enough to pay for it with the precious currency of his social acceptance?"

History seems to bear this thesis out. With the benefit of hindsight we can see that in every generation the test has been especially rigorous at one or two chosen points. At such points, seekers were required to break with a majority of many on earth in order to please a minority of One in Heaven. In medieval Europe, for example, it was not at all costly to believe that the god of the Bible created the cosmos. It was, however, very costly (and, we now reckon, heroic) to challenge certain Roman Catholic interpretations of the Bible. Today it is not at all costly to believe that the universe evolved from an exploding singularity. It is, however, very costly to challenge that view in public—and costlier still to affirm that the biblical view is true after all. Could it be, then, that in our own time the unknown god has chosen the beginning as a special point of human testing? If so, it will take great courage to pass.

The second challenge, however, is even more daunting than the first. For here the test perspective will again require you to stand alone, but this time before the god of the Bible. And here, in the eyes of many, is the summit of human testing. If they are right, you will need your every ounce of honesty, courage, and love of the truth to reach the top.

Consider again the biblical beginning. You now know it must be reckoned with, since, as we have just seen, numerous trails of good evidence all converge in this single clearing. And yet you are tempted to flee it. Why? *Because you know, deep down, that it threatens to precipitate a definite spiritual crisis should you come to believe that it is true.*

You know, for example, that you cannot stop at Genesis 1 and 2, but must go on to Genesis 3.

You know you cannot rejoice in Elohim (the wise, powerful, and loving God of creation), yet deny Yahweh (the holy and righteous God who governs, tests, rewards, and judges his people).

And you know you cannot be a child of the holy God without also being a child of the sinful Adam.

Accordingly, you find this cosmology both encouraging and unsettling. It beckons you not only to meet your maker, but also to bow before your ruler and your judge. It offers you hope, but also demands change. It invites you into a new world, but requires you to repudiate and leave behind the old. In short, it speaks to you as to an autonomous human self—a self that does not want to submit to a Self higher than itself—and bids you come and die.

If you doubt all this, please listen carefully to the following words of Dr. George Wald, professor emeritus of biology at Harvard, as he explains his devotion to evolutionism with alarming candor:

> There are only two possible explanations as to how life arose: spontaneous generation arising to evolution, or a supernatural creative act of God... There is no other possibility. Spontaneous generation was scientifically disproved 120 years ago by Louis Pasteur and others, but that just leaves us with only one other possibility... that life came as a supernatural act of creation by God. *But I can't accept that philosophy because I do not want to believe in God.* Therefore, I choose to believe in that which I know is scientifically impossible—spontaneous generation leading to evolution.[1]

Listen also to evolutionary biologist Dr. Richard Lewontin, who says much the same thing in these oft-cited words:

> We take the side of science *in spite* of the patent absurdity of some of its constructs, *in spite* of its failure to fulfill many of its extravagant promises of health and life, *in spite* of the tolerance of the scientific community for unsubstantiated just-so stories, because we have a prior commitment, *a commitment to materialism.* It is not that the methods and institutions of science somehow compel us to accept a material explanation of the phenomenal world, but, on the contrary, that we are forced by our *a priori* adherence to material causes to create an apparatus of investigation and a set of concepts that produce material explanations, no matter how counter-intuitive, no matter how mystifying to the uninitiated. Moreover, that materialism is an absolute, for we cannot allow a Divine Foot in the door.[2]

These quotations are as instructive as they are troubling. They teach us that natural science is not the bastion of objectivity that we imagined it to be. They teach us that some people accept cosmic evolution, not because it is reasonable or evidentially sound but because it offers a way of escape from the unknown god and from certain unwelcome changes that he might require them to make. But most importantly, they teach us that there is something *in us all* (and not just in those materialists out there) that values personal autonomy above truth, self above god; something that does not want us to die, even if such a death were the very gateway to eternal life.

To stand alone before the god of the Bible is, I have found, to meet that something. To get past it and into the truth will take all the courage you have, and more besides. But to do so is to triumph in the test of life.

And that, dear friend, is the way to the beginning, as best I see it. Before you go, let me sum it up, that you may ever keep it in mind.

Embrace the test perspective. Have faith in the goodness of the unknown god. Seek his truth diligently, with all your heart. Trust your god-given faculties, no matter how feeble they may seem. Above all, be very courageous: courageous enough to stand alone before man, and courageous enough to stand alone before the god of the Bible.

If you do these things the unknown god will surely take note. Seeing your love of the truth, he will come to you in the way and introduce you to his Teacher, who will take you back, back—all the way back to—the beginning.

When you arrive, be sure to get in touch. Like children, we will rejoice together in all that you have seen.

NOTES

1. George Wald, "Origin, Life and Evolution," *Scientific American*, 1978.

2. Richard Lewontin, "Billions and Billions of Demons," *The New York Review* (January 9, 1997), p. 31.

The Unity of the Bible

Many believe that the unity of the Bible is the preeminent proof of its divine inspiration, authority, and trustworthiness. But what exactly is meant by the term "unity," and how does the Bible display this telling characteristic?

As we saw in our study, the phenomenon of unity is inseparable from *order*. We cannot behold unity unless we see it in an order of some kind. An order may be defined as *a collection of component parts that has been integrated (i.e., unified) into a system by means of a definite plan*. A strand of DNA, a cell, a flower, an eye, an ear, a brain—all are examples of naturally occurring orders. They are collections of component parts integrated into fantastically complex, beautiful, and functional systems according to a definite plan. Just to look at them is to know these orders could not possibly have arisen by accident. Self-evidently, they require and reveal a divine Orderer. They are one of the great proofs for the existence of a rational, powerful supreme being—a divine creator and preserver—who is at work in the natural world.

The Bible too is an order. Like an object in the fog, its orderliness requires some forward momentum on our part to be seen clearly—some time and study. But if we are willing to exert ourselves, we realize soon enough that the Bible does indeed have many component parts, and that they too are woven into a fantastically complex, beautiful, and functional system by means of a rational plan. Indeed, over time—and by God's gracious work of illumination—this unity will not only become evident, but compelling. It will be impossible to view the Bible as a random collection of Jewish myths and musings. No, it must be the purposeful creation of a rational supreme being—a divine revealer of truth—who seeks to work in the minds of men and nations. In short, we will come to see the Bible as a very special gift from the unknown god, a gift in which he discloses to all mankind the much-needed answers to the questions of life.

The outline below, summarizing material found in the Introduction to this book, is designed to display concisely the architecture and implications of the unity of the Bible. May it inspire you to further study of the Bible itself!

THE UNITY OF THE BIBLE

I. **MULTIPLICITY**

    A. Sixty-six different books

    B. Written in three different languages (Hebrew, Aramaic, Greek)

    C. In eight different literary genres

    D. By about forty different authors

    E. Over the space of about 1600 years (ca. 1500 BC to 70 AD)

    F. Concerning thousands of persons, places, things, and events

II. **UNITY**

    A. One story (the creation, fall, and redemption of man and the cosmos)

    B. About one God (the triune Yahweh; Father, Son and Holy Spirit)

C. Administering one plan of salvation (an eternal covenant between God and man, concealed in the Old Testament, revealed in the New)

D. Centered around one Person (the redeemer, Jesus Christ: prophet, priest, and king)

E. Who is attested to by one (large and diverse) body of signs:

1. Signs surrounding Jesus' birth

2. Angelic visitations and testimonies

3. Theophanies

4. Miracles

5. The Resurrection

6. OT Messianic types

7. OT Messianic prophecies

8. The Church

F. And worshipped by one people (believing Jews and Gentiles, past, present, and future)

G. According to one (clear and comprehensive) worldview (biblical answers to the questions of life)

III. IMPLICATIONS: ITS UNITY IMPLIES THAT THE BIBLE IS (OR WILL BE)...

A. Divinely inspired (given by God through inspired men, 2 Tim. 3:16-17).

B. Inerrant (true in all it affirms, John 10:35, 17:17).

C. Complete (no more scriptural revelations to come, Eph. 2:19-22, Jude 1:3, Rev. 22:18-19).

D. Trustworthy (Matthew 7:24-28).

E. Authoritative (Matthew 7:29).

F. Preserved (Matthew 24:36).

G. Recognized (as God's Word, Luke 24:45, 1 Thessalonians 2:13).

H. Infallible (it will certainly accomplish what it was sent to do, Isaiah 55:11, Colossian 1:3-6).

Old Testament Messianic Types

In a tense confrontation with the religious leaders of his day, Jesus of Nazareth challenged his opponents, saying, "You search the Scriptures, for in them you think you have eternal life; and these are they that testify of me" (John 5:39). Here Jesus daringly claims that the Jewish Scriptures cannot stand alone; that by their very nature they are forward-looking revelations; that they await the coming of someone who will fulfill them and thereby unveil their hidden meaning; and that he himself is the one who does these very things. Thus Jesus might well have said here what he had said earlier to his own disciples: "Do not think that I have come to destroy the Law and the Prophets. I have not come to destroy, but to fulfill" (Matthew 5:17).

Truly, these were radical statements. By introducing a new motif of promise and fulfillment, Jesus was actually introducing a new way of interpreting the Jewish Scriptures. Henceforth, he implied, men must see these writings as the record of life under an "old testament"— an old covenant (or agreement) that was secretly preparing the way for a new (Jeremiah 31:31f, Matthew 9:17, Luke 22:20, Hebrews 8:8-12). Moreover, they must now interpret this Old Testament (OT) "christo-centrically"—as mysteriously pointing ahead to the person and work of the Messiah, Jesus Christ (Luke 24:44-49). In reading its history men must therefore ask themselves "What is the hidden, Messianic significance of these stories? What do they teach me about the Christ who was yet to come and in whose appearing they are now fulfilled at last?"

Jesus' apostles definitely got the message. In their writings both to Jew and Gentile, we find them repeatedly citing or referring to the OT scriptures, seeking to display Christ in them all. Indeed, they even developed a special vocabulary to speak about this new method of interpretation, declaring that upon the vast terrain of OT history God has strewn, like so many precious gems, innumerable *types* (Greek, *tupos*: figure, symbol) of Jesus Christ (Romans 5:14, 1 Corinthians. 10:1-13, Hebrews 8:1-6, 1 Peter 3:21). A Messianic type, they said, is any OT person, place, thing, institution, or event which prefigures Jesus of Nazareth, the events of his life, and the blessings of the new covenant he came to bring. The OT is loaded with them. They are, as it were, a vast multitude of shadows, all cast by one body: Christ and the things of his covenant (Colossians 2:17). With the help of the Holy Spirit, people can spot them and see how they have been (or yet will be) fulfilled in Christ (Luke 24:45, 2 Corinthians 3). Thus, to understand the Old Testament at its deepest level, one must interpret it *typologically*, in terms of Jesus Christ.

Together with OT Messianic prophecies, the Messianic types have traditionally stood among the preeminent proofs of the divine inspiration of the Bible and the spiritual authority of Jesus of Nazareth. Today, however, many who have heard about Messianic prophecy know little or nothing about Messianic types. This is unfortunate, since the types are, if possible, even more impressive than the prophecies. In the prophecies God speaks of the coming Messiah through men (the prophets), showing that he knows the future. In the types, however, he speaks of the Messiah through historical events, showing that he not only knows the future but creates it as well! The types display Israel's God as the sovereign controller of all history, even down to its minutest details!

The table below is designed to introduce seekers to the amazing but oft-neglected world of Messianic types. I offer it as an inducement for you to study the Bible and to determine for yourself whether or not the OT really can be interpreted typologically and christo-centrically. This will take some time and effort. In particular, it will require you to familiarize yourself with OT history (Genesis through Esther), the Gospels and the Epistles. If you are new to the Bible, I recommend that you read the Gospels first, then the OT historical books, then the Epistles. As you travel back and forth upon the great plain of biblical history, keep your eyes opened wide. You too may see what multitudes of others have seen—astonishing correspondences between OT history and the things of Christ; correspondences so intricate, so beautiful, so numerous, and so improbable as to create a single indelible impression: Jesus of Nazareth must indeed be the Teacher come from God, and the Bible his inspired Word.

OLD TESTAMENT MESSIANIC TYPES			
OT Ref	Type	Fulfillment	NT Ref
Gen. 2	Adam, head and progenitor of an earthly race	Christ, head and progenitor of a heavenly race	Rom. 5ff I Cor. 15:20-28
Gen 2-3	The Tree of Life	Christ, in whom is eternal life. The cross, where he died to give life to God's people	Jn. 5:21, 10:10 Rev. 2:7 Gal. 3:13
Gen 6-9	The Flood, Noah and the ark	Christ, the ark of God, safely delivering his people through fires of judgment to a new heaven & a new earth	I Pet. 3 2 Pet. 3
Gen. 14	Mekchizedek, high priest of Salem	Christ, eternal high priest of his people, "the Jerusalem above"	Heb. 5, 7, 12
Gen. 22, 24	Abraham offers his only son on wood as a sacrifice to God	God the Father offers his only Son on wood as a sacrifice to himself—and receives him back alive from the dead	Heb. 11
Gen. 28	Jacob's Ladder	Christ, the one mediator between Heaven & earth, God & man	Jn. 1:51 I Tim. 2:5-6
Ex.-Josh.	Moses delivers Israel from Egypt and leads them through the wilderness to Canaan	Christ delivers his people from bondage to evil and leads them through the fallen world-system to a new heaven & a new earth	Gal. 1:3-5 I Cor. 10 Heb. 4, 11 Rev. 12, 20
Ex. 12	The Passover Lamb	Christ, the lamb of God, slain for his people so that he may pass over them in the day of judgment	Mat. 26 Jn. 1, 19 I Cor. 5:6-8
Ex. 16	Manna, physical food for Israel in the wilderness	Christ, the bread of heaven, spiritual food for God's pilgrim people in the world	Jn. 6 Rev. 12
Ex. 17	Moses strikes the rock at Horeb, supplying water for Israel	God strikes Christ once for all, to supply the water of life to his people	Jn. 4, 7; Acts 2 I Cor. 10; Rom. 6 Heb. 7
Ex. 25	The mercy-seat	Christ, the place of meeting between God & man	Rom. 3, 5; Col. 1 Heb. 4:14-16
Lev. 16	The scapegoat	Christ, the bearer and remover of his people's sins	Heb. 10
Jonah 2-3	Jonah, a reluctant prophet swallowed by a sea monster	Christ, a willing prophet, swallowed by death but recalled to life by God to preach good news	Mt. 12:38-40

Appendix 3

New Testament References to Genesis 1-11

Like all orthodox Jews, Jesus and his apostles assumed that Genesis 1-11 was a divinely inspired account of the good and bad beginning. They regarded it, not as myth or poetry, but as true history. They found instruction, warning, and comfort in the persons, places, things, and events about which Moses wrote. Indeed, a careful study of the passages listed below will show that Genesis 1-11 was foundational to the apostolic worldview. In particular, all of the NT writers presupposed that the creation and fall of Genesis 1-3 set the stage for a divine redemption that was prepared for in OT times and was now being fulfilled in their own.

The importance of Genesis 1-11 is indicated by the frequency with which the NT authors refer to it. Dr. Walter Brown, in an exhaustive survey, finds sixty-eight such citations or references. He also notes that every NT author refers to Genesis 1-11, that Jesus of Nazareth referred to each of the first seven chapters, that the NT authors refer to ten of the eleven chapters, and that sixteen out of twenty-seven NT books refer to Genesis 1-11.

The table below is a much-abbreviated version of Dr. Brown's, spotlighting some of the most important NT references to Genesis 1-11. For a complete listing, see Dr. Brown's excellent book, *In the Beginning*, (CSC Publications, 2001), pp. 283-4.

NEW TESTAMENT REFERENCES TO GENESIS 1-11		
NT Ref	Subject	Genesis Ref
Mt. 23:35	The death of righteous Abel	4:4
Mt. 24:37-9	The dark days of Noah	6:1-8:22
Mk. 10:6-8	Marriage ordained at the beginning of the creation	1:27, 2:24, 5:2
Lk. 3:23-38	Jesus' genealogy traced through Adam to God	5, 10, 11
Jn. 8:44	Satan, father of lies	3:4-5
Rm. 1:20	Man present from the creation of the world	1, 2
Rm. 5:12-21	Adam, earthly head of the human race, through whom sin entered the world	2:15-17, 3:1-19
Rm. 8:20-22	Nature cursed because of Adam's sin	3:17-19
1 Cor. 11:2-16	Man ordained head over woman at creation	1:27, 2:18, 22-23, 3:16
1 Cor. 15:21-2	Death came through Adam	2:16-17, 3:19
1 Cor. 15:45-9	Adam, made of dust, prototype of "natural" men and women	2:27, 3:23
Eph. 5:30-31	Man must cleave to his wife, become one flesh with her	2:24
1 Tim. 2:13-14	Adam created first; Eve deceived	2:18-23, 3:1-6, 13
Heb. 4:1-10	God's Sabbath rest	2:2-3
Heb. 11:1-7	Faith's Hall of Fame: Abel, Enoch, Noah and his household	4:3-5, 5:21-24, 7:1
1 Pet. 3…	Noah, a type of all saved through water (baptism)	6:14-16, 7:13
2 Pet. 3:4-5	Heaven and earth created by the word of God	1:1-2:3
1 Jn. 3:12	Jealous Cain, prototype of worldly persecutors	4:8, 25
Jude 14	Enoch, prophet of judgment	5:18-24
Rev. 2:7, 22:2, 14	The Tree of Life, regained by God's redeemed children	2:9

Select Bibliography

Recent Creation

Batten, Don, *The Revised and Expanded Answers Book* (Master Books)

Brown, Walt, *In the Beginning: Compelling Evidence for Creation and Flood* (CSC Books)

Byl, John, *God and Cosmos: A Christian View of Time, Space and the Universe* (Banner of Truth)

Kelly, Douglas, *Creation and Change* (Mentor Books)

Morris, Henry and John, *The Modern Creation Trilogy* (Master Books)

Morris, Henry, *The Biblical Basis of Modern Science* (Baker Books)

Sarfati, Jonathan, *Refuting Compromise* (Master Books)

_____, *Refuting Evolution* (2 Volumes) (Master Books)

Intelligent Design

Behe, M., *Darwin's Black Box* (Simon and Schuster)

_____, *Science and Evidence for Design in the Universe* (Ignatius)

Denton, M., *Evolution: A Theory in Crisis* (Adler and Adler)

Dembski, W., *Signs of Intelligence: Understanding Intelligent Design* (Baker)

_____, *The Design Revolution: Answering the Toughest Questions About Intelligent Design* (InterVarsity)

_____, *Uncommon Dissent: Intellectuals Who Find Darwinism Unconvincing* (Isi Books)

Strobel, L., *The Case for the Creator* (Zondervan)

Wells, Jonathan, *Icons of Evolution: Science or Myth? Why Much of What We Teach About Evolution is Wrong* (Regnery)

To order additional copies of

IN SEARCH OF
THE
BEGINNING

Have your credit card ready and call:

1-877-421-READ (7323)

or please visit our web site at
www.pleasantword.com

Also available at: www.amazon.com
and
www.barnesandnoble.com

Printed in the United States
68941LVS00004B/118

9 781414 103716